History of Mechanism and Machine Science

Volume 48

This bookseries establishes a well-defined forum for Monographs and Proceedings on the History of Mechanism and Machine Science (MMS). The series publishes works that give an overview of the historical developments, from the earliest times up to and including the recent past, of MMS in all its technical aspects.

This technical approach is an essential characteristic of the series. By discussing technical details and formulations and even reformulating those in terms of modern formalisms the possibility is created not only to track the historical technical developments but also to use past experiences in technical teaching and research today. In order to do so, the emphasis must be on technical aspects rather than a purely historical focus, although the latter has its place too.

Furthermore, the series will consider the republication of out-of-print older works with English translation and comments.

The book series is intended to collect technical views on historical developments of the broad field of MMS in a unique frame that can be seen in its totality as an Encyclopaedia of the History of MMS but with the additional purpose of archiving and teaching the History of MMS. Therefore. the book series is intended not only for researchers of the History of Engineering but also for professionals and students who are interested in obtaining a clear perspective of the past for their future technical works. The books will be written in general by engineers but not only for engineers. The series is promoted under the auspices of International Federation for the Promotion of Mechanism and Machine Science (IFToMM).

Prospective authors and editors can contact Mr. Pierpaolo Riva (publishing editor, Springer) at: pierpaolo.riva@springer.com.

Indexed by SCOPUS and Google Scholar.

Giuseppe Genchi

The Museum of Engines and Mechanisms of the University of Palermo

The Collection Catalogue

Springer

Giuseppe Genchi
Museum System of the University
of Palermo
Palermo, Italy

ISSN 1875-3442 ISSN 1875-3426 (electronic)
History of Mechanism and Machine Science
ISBN 978-3-031-63041-5 ISBN 978-3-031-63042-2 (eBook)
https://doi.org/10.1007/978-3-031-63042-2

This Springer imprint is published by the registered company Springer Nature Switzerland AG
The registered company address is: Gewerbestrasse 11, 6330 Cham, Switzerland

On ne connaît pas complètement une science tant qu'on n'en sait pas l'histoire

—Auguste Comte

Foreword by Massimo Midiri

The University of Palermo is one of the main players in the social, economic and cultural progress of Sicily, to which it contributes through its Research, Teaching and Third Mission activities. The latter is dedicated to the transfer of knowledge to the community and characterizes the role of cultural developer of the University. A significant part of these activities is closely connected with the vast and heterogeneous historical heritage of our university, made up of artistic, naturalistic and technological assets. Acquired in over two centuries of institutional activity, nowadays they overall represent the history and the academic tradition of the University of Palermo, defining its cultural identity. The protection and enhancement of this heritage is entrusted to the University Museum System, which plays a remarkable role in Third Mission accomplishment through multiple relationships and cultural cooperation with the territory. These derive primarily from all the activities aimed at making the historical heritage of the University increasingly accessible to the students as well as the entire community. All this is well represented by the Museum of Engines and Mechanisms, which is the youngest of the six museums of the University of Palermo. Inaugurated in 2011, this Museum was made up thanks to the far-sighted vision and perseverance of Giuseppe Genchi, who, starting from 2008, transformed an amorphous set of finds to be scrapped into a well-established museum structure organized, equipped with a vast scientific collection, restored, protected and, above all, made accessible to the community. His continuous research, restoration and cultural promotion activities have led to a notable development of the collection, now almost unique in its kind, and, at the same time, have allowed the Museum to fully establish itself in the field of industrial heritage, as demonstrated by important international awards received in recent years. As a result, nowadays the Museum is a dynamic structure in step with the times, like its collection itself, which represents the historical and technical evolution of engines and a large part of applied mechanics, highlighting their connections with many technological, economic and social topics. A historical and cultural heritage that is continuously enhanced by means of new purchases and donations as well as to various cooperation projects developed with other institutions and companies, which recognize the excellence of the Museum by entrusting it with the protection and enhancement of their historical heritage.

April 2024

Massimo Midiri
Rector of the University of Palermo
Palermo, Italy

Foreword by Michelangelo Gruttadauria

The Museum System of the University of Palermo is a widespread museum composed of six scientific museums, numerous collections and historical sites which represent a very broad and complete view of knowledge, ranging from classic art to natural sciences, medicine and engineering. Thanks to the constant dedication, skills and professional experience of the staff and collaborators of the Museum System, the historical heritage of the University is alive and continuously enhanced, through restoration activities and cultural events, aimed at the wider public engagement. To this purpose, the Museum System carries out a great work of opening, connection and integration between the academic world and the society within the development policy promoted by the University. This in turn characterizes the Museum System strategy for what concerns the management and the activities of its museums, which are at the same time places of research and teaching as well as spaces of cultural dissemination and public enjoyment. This gives academic museums a particular charm but also requires the continuous search for a formal and substantial balance between the academic thoroughness and the need for an easy and comprehensible communication, in step with the times. This peculiar approach leads many educational and scientific activities such as the publication of articles, books and the participation in scientific conferences. All characteristics that are well represented in this catalogue, aimed at illustrating one of the most noteworthy historical collection of engines and mechanisms through the scientific and philological method for technical explanation of design features of each collection specimen together with a description of the historical context.

April 2024

Michelangelo Gruttadauria
President of the Museum System
of the University of Palermo
Palermo, Italy

Foreword by Marco Cammalleri

The preservation of the historic heritage is one of the priorities that a civil society must set itself. This principle does not only concern works of literature, art, architecture, but also works of a technical-scientific nature.

Starting from the 19th century, the development of thermodynamics, which had already started empirically in the second half of the previous century, allowed the creation of new machines capable of carrying out, with ever greater efficiency, those chemical-physical transformations necessary to obtain levels of mechanical power previously unimaginable. For this reason, steam engines played a decisive role in the Industrial Revolution in which, as in a chain reaction, the development and use of driving and operating machines spread everywhere, from the manufacturing sector to the transport sector. In general, this has led, gradually but irreversibly, to important economic and social transformations, with the consequent transition from a purely rural society to a modern, industrial and technological one. Even today, more than a century later, in the era of great digital innovations, machines, understood in a general sense as complex devices capable of transforming energy, remain, beyond the specific scope of application, the basis of modern technological society and are directly or indirectly connected to numerous activities of our daily life. Understanding its history and development prospects therefore means enriching one's scientific and cultural heritage, but not only; studying the history of technology can also be a source of inspiration and help engineers be more creative designers. In fact, technological progress is hardly achieved through revolutionary inventions but rather by the accumulation of successive improvements over time. Many solutions that appear innovative today are reinterpretations of others conceived in the past and then abandoned, or at the time insurmountable technological difficulties, or because they were conceived for an application field different from the current one. All this gives the study of the history of technology and technological museums, especially academic ones, an additional value to the obvious one of a purely cultural nature. An important aspect that characterizes the Museum's initiatives is therefore represented by the active involvement of young people, especially students, in the realization of multiple educational and cultural activities, including guided tours of the collection focused on technical-scientific themes, seminars with important speakers from the academic world and the industrial sector, training internships

Fig. 1 The collection at the beginning of the restoration project in April 2008

in the context of various study courses, not only in the engineering field. Some initiatives are dedicated to schools and are partly realized thanks to the collaboration relationships developed with various educational institutes. The training activities also include various projects with the National Civil Service, of which the Museum is an accredited headquarters. Each project allows young people to join the Museum staff for an entire year and includes both training activities and practical support activities in the various operational areas of the Museum.

From all this derives the importance of preserving, making usable and enhancing the collections of machinery and technical equipment, as precious evidence of the history of scientific progress and of the various educational, historical, cultural aspects.

The Museum houses a vast collection of engines and industrial, scientific and educational equipment, acquired and used in various research and teaching sectors starting from the second half of the 19th century, with the establishment in 1866 of the Royal Application School for Engineers and Architects at the University of Palermo. All these finds constitute a heritage of great value that describes the evolution of fluid machinery as well as the studies connected to it, in many fields of practical application. The Museum's collection (Figs. 1 and 2) includes various engines from the end of the 19th century, both steam and internal combustion type, many automobile and motorcycle engines, which stand out for their rarity and value, from that of an Isotta Fraschini 8A to the contemporary 12-cylinder of the Lamborghini Aventador.

Fig. 2 One of the exhibit areas of the Museum, in the same place showed in Fig. 1

Fig. 3 An overview of the first part of the aircraft engines exhibit

A significant part of the collection is made up of the numerous aircraft engines (Fig. 3), among which there are some unique and other very rare examples, from the rotary radial engines of the early aviation period to modern turbojet engines. The Museum's collection also includes one of only five complete surviving FIAT G. 59 aircraft from 1952, one of the last high-performance aircraft equipped with a piston engine. An entire section is dedicated to marine engines, recently expanded thanks to the acquisition of some FIAT Grandi Motori engines. This includes particularly rare finds, ranging from a 19th-century steam engine to large supercharged two-stroke Diesel engines. Among these, one stands out for its size, rarity and value, the FIAT 2 C. 116 engine from 1909 that equipped the "Medusa" submarine of the Italian Royal Navy, which sunk off the coast of Venice during the First World War.

Fig. 4 An overview of a part of the stationary engines exhibit

The museum exhibit layout is arranged in a simple and linear way: the collection elements are grouped in specific areas according to the type of application (automotive, aviation, naval, etc.), in which the items are displayed in chronological order to illustrate their technical evolution (Figs. 4 and 5). Each engine is endowed of a large descriptive panel containing the main technical and historical characteristics, together with photographs and explanatory drawings. Given the academic and didactic nature of this Museum, the vast iconographic kit is structured according to a technical-scientific setting, while maintaining broad accessibility to the public of all kinds through different levels of detail.

All the information is made freely available both on the Museum's website, both in English and Italian language, as well as on other more interactive cultural dissemination and promotion platforms, such as the audio-guides of the izi.TRAVEL portal. Furthermore, the Museum is a partner of the *Once Upon a Try* international program promoted by Google Arts and Culture, in the context of which the entire collection is made freely accessible, including a complete 3D virtual tour of the exhibit areas, realized by Google Street View technology.

That prestigious program, aimed at disseminating the history of technical and scientific development, today represents one of the most important digital exhibit showcases and is created through the non-profit contribution of an international selection of 120 museums, scientific and industrial institutions, including NASA, Smithsonian National Air and Space Museum and Deutsches Museum.

From a technical point of view, the Museum has a high producing capability since it takes care of all the restoration activities internally in its own laboratory, in which all the collection items are treated following the canons of philological restoration and through a modern conservative approach, where it is possible. Based on the needs of each project, traditional restoration methods are often integrated with more modern techniques derived from other technical-scientific sectors. For the implementation

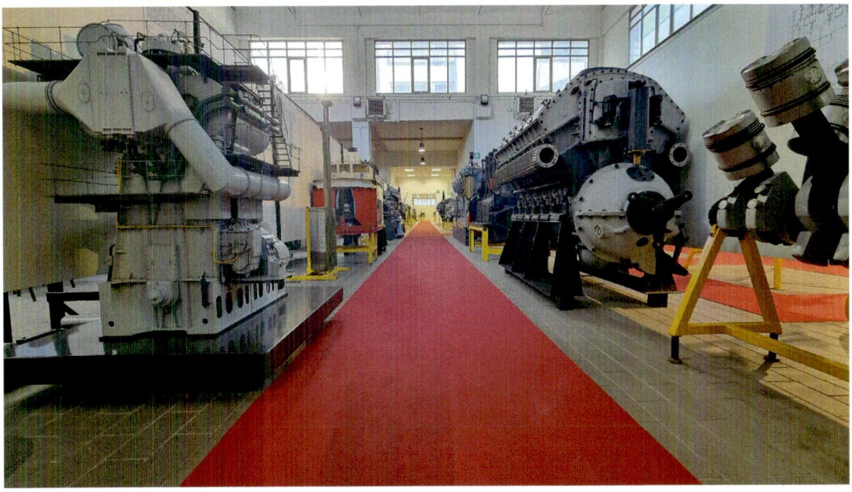

Fig. 5 An overview of the marine engines exhibit

of some of the more complex activities, the Museum makes use of the technical and scientific support of the Engineering Department of the University of Palermo or makes use of specialized companies within its network of external partners. The Museum's laboratory also provides for all the periodic and extraordinary maintenance operations of the collection items and technical equipment, as well as for the construction of the specific support of each engine and all needs for new exhibits.

Due to the importance of its collection and as a result of many restoration activities carried out in recent years, the Museum received—the first and currently the only one in Italy—the prestigious international award Mechanical Engineering Heritage Collection assigned by the American Society of Mechanical Engineers (ASME), as part of the History and Heritage Landmarks program aimed at enhancing and promoting the most important historic mechanical heritage around the world.

The Museum is the youngest among those of the Museum System of the University of Palermo. It was inaugurated in 2011 but its story begins in 2008, when a young mechanical engineering student, peering through the windows of a warehouse in the mechanical workshop, glimpsed an indistinct set of engines and parts (Fig. 1), scattered and piled up at the same time, in the process of final disposal.

The extraordinary passion and expertise for historic motoring, however, allowed him to understand that a real treasure was hidden in that pile of wreckage. He then managed to convince his Applied Mechanics professor, Riccardo Monastero, and the senior technician Beniamino Drago to allow him to start historic research, which confirmed the potential value of those piled up "pieces". Hence his idea, which seemed foolish and visionary to most, was to preserve that heritage and make it a museum. He then volunteered to restore an engine with the initial help of Mr. Drago; then another and another. He worked intensely for years, in his free time and on weekends, combining the manual skills of the restoration with increasingly in-depth

Fig. 6 The restoration of the Isotta Fraschini 8A engine operated at the laboratory of the Museum by means of both traditional and modern methods, such as reverse engineering processes

Fig. 7 The restoration of the historic aircraft FIAT G. 59, during its final assembly in a dedicated area of the Museum, in which is nowadays publicly displayed

historical research of what would gradually turn out to be real finds of enormous historical and collectible value (Figs. 6 and 7).

On 25th February 2011, the collection of engines and mechanisms restored up to that date received the recognition of Academic Museum, Prof. Riccardo Monastero was appointed director, and the student who had invented and created it was awarded the official commendation by the Rector of the University of Palermo, Prof. Roberto Lagalla. From that day, the Museum gradually expanded with the addition of many new items such as marine engines, as well as instruments, mechanical parts and historic documents. This nucleation and growth process has strengthened over time, thanks to the research and restoration activities carried out on the finds already in storage and on those acquired through private donations, transfers on loan from other entities, both public and private, including large international companies such as Stellantis, Toyota, Ford, Ferrari and Lamborghini. In this context, one of the most important projects was launched in 2021, in collaboration with the Italian Air Force, which entrusted over 40 historic engines to the Museum for their restoration and public display. Due to all the above-mentioned activities, carried out systematically since 2008, the number of engines displayed at the Museum was doubled in just over ten years and the collection of parts, models and technical equipment was considerably increased.

Fig. 8 The Museum's depository preserves various engines awaiting their restoration

Nowadays the young student protagonist of this story is Giuseppe Genchi, Ph.D., Conservator of the Museum and author of this book. All my personal and institutional acknowledgment goes to him for having given birth to this Museum, seeing where others could not see, and for having been subsequently, over the years, promoter and "engine" of every activity that allowed this Museum to become today, in its particularity and uniqueness, the most important and heterogeneous collection of engines in Italy, as well as one of the most significant in Europe.

June 2023

Marco Cammalleri
Scientific Director of the Museum
of Engines and Mechanisms
Palermo, Italy

Preface

This catalogue is the first complete publication, in English language, entirely dedicated to the description of the main items of the collection of the Museum of Engines and Mechanisms of the University of Palermo. Its heritage is one of the most important and heterogeneous collection of engines and mechanisms, among those freely available in Europe, the first in Italy to be awarded Mechanical Engineering Heritage Collection by the American Society of Mechanical Engineers. Thanks to its rare and valuable items, this unique collection shows the evolution of fluid machinery, from steam engines up to turbojet engines, as well as hybrid system. With over 150 engines, numerous didactic models, parts and technical equipment, including various unique exemplars, the collection is continuously enhanced thanks to donations and through restoration activities carried out in a dedicated laboratory of the Museum.

Each of the main fields of application of the engines and other collection elements, such as the two aircraft, is well displayed in a specific exhibit area, in which the variety of items makes it possible to represent all the main technical solutions developed over the time. Given the complexity of the evolutionary path of motoring, closely connected with various scientific, historical and economic aspects, the structure of the catalogue takes up that of the exhibition layout inside the Museum, based on the subdivision into categories relating to the aforementioned areas of practical application. In each of these, the finds follow one another in chronological order. This is only one of the possible schematizations but, more than others, it facilitates the understanding of the development path of the items, as well as the relative contextualization from the technical, historic and collecting point of view.

This catalogue is the synthesis of numerous historical-technical research campaigns carried out in the last 14 years, from the beginning of the restoration project of the collection and the contemporary set up of the Museum (Fig. 8). The history of these activities is also linked to the systematic expansion of the collection and has been full of unexpected developments, connections with other cultural fields and collaborations with other researchers, which have led to the creation of new research projects, the acquisition of new items, publications, cultural and educational activities of the Museum. The research activities, carried out starting from the primary sources and with a philological approach, were necessary and useful for rigorously conducting the technical restoration activities, the cataloguing files, the creation of various publications and the vast iconographic kit that enriches and characterizes the museum layout. A lot of information has been found thanks to the support and collaboration of important organizations and companies that have allowed access to their historical archives. Among these, the most important was the FIAT Historic Centre in Turin, which, thanks to its vast heritage, represents one of the main points of reference for the history of motoring in Italy.

As a result of a great deal of research and restoration activities carried out since 2008, this book constitutes the identity document of the collection and, more generally, of the Museum itself. Despite the technical subject and the academic environment in which it was created, the catalogue is realized to be read even by non-experts by means of different levels of detail, the first of which is the historical and economic context related to an engine, as well as to the vehicle for which it was designed and used. Since the space in the catalogue is limited compared to the size of the collection and the amount of data available, only the most important items are described in detail with a dedicated paragraph. Each one contains a large color picture of each item, followed by a general description of its history, a summary table of the main technical specifications and, where possible, some original technical drawings, that clearly illustrate the internal structure and some specific features.

Finally, in the last two years the growth rate of the items acquired by the Museum has certainly been faster than the study and research activities which are the basis of the Museum's scientific production. Many of the items recently acquired, briefly mentioned in the appendix in this catalogue, will be the object of the next development activities of the Museum.

Giuseppe Genchi
Founder and Curator of the Museum
of Engines and Mechanisms

Palermo, Italy
June 2023

Fig. 9 The rare counter-rotary aircraft engine Siemens Halske Sh.IIIa

Acknowledgments

Massimo Midiri, Michelangelo Gruttadauria, Marco Cammalleri, Beniamino Drago, Claudia Lo Re, Floriana Dolce, Rosario Seidita, Giulia Andronico, Giorgio Hardouin, Angelo De Stefani, Lorenzo Morello, Luca Manneschi, Basilio Di Martino, Giovanni Francesco Adamo, Luca Carapellese, Marco Ceccarelli, Lorenzo Ardizio, Maurizio Torchio.

Abstract

The book describes the collection of the Museum Engines and Mechanisms of the University of Palermo, Italy, one of the most important and heterogeneous collections of engines and mechanisms in Europe, the first one in Italy to be awarded as Mechanical Engineering Heritage Collection by the American Society of Mechanical Engineers. Thanks to its numerous items, this book showcases the evolution of fluid machinery and applied mechanics, from steam engines up to turbojet engines, as well as hybrid system, giving several technical and historical information about its most important engines, which are described in detail through pictures and original drawings. The Museum preserves and makes freely available this almost unique collection of more than 300 engines, didactic models, and technical equipment, including various unique exemplars, continuously enhanced thanks to donations and through restoration activities carried out in a dedicated laboratory of the Museum. As a result of a great deal of philological research carried out on the documents collected in the Museum's archive, as well as in other institutional and corporate historical archives, this book serves as the reference tool of the collection and, more generally, of the Museum itself. Despite the technical subject and the academic environment in which it was created, the catalogue is realized to be read even by non-experts, offering different levels of detail, the first of which is the historical, economic and, in certain cases, even sporting context related to an engine, such as the vehicle for which it was designed and used.

Contents

Chapter 1
Stationary Engines

Abstract Steam engines were the first modern machines used for energy transformation, playing a crucial role in the First Industrial Revolution in the eighteenth century. Their use in manufacturing, transportation, and electricity production led to the transition from the agricultural economy to the modern and industrial one. In that period was founded the Royal Application School for Engineers and Architects in Palermo in which the older engines of the Museum collection were operated for scientific and teaching purposes, such as the Neville stationary engine, which well represents the typical steam engine widely used up to early'900 in textile factories, mechanical industries, and mines. The museum also houses a rare Willans compound high speed engine and a Ljungström steam turbine from the 1920s, originally used in a powerplant of Palermo up to 1952. The collection also includes several internal combustion engines from an early Langen & Wolf engine dating to the 1880s, up to a Perkins turboshaft. Among stationary engines, various hydraulic machines, such as Francis, Kaplan, and Pelton turbines, are also displayed, in order to highlight the first evolution of the fluid machinery technology throughout history.

G. Genchi, *The Museum of Engines and Mechanisms of the University of Palermo*,
History of Mechanism and Machine Science 48,
https://doi.org/10.1007/978-3-031-63042-2_1

Fig. 1.1 The exhibit area of the Museum dedicated to stationary engines also includes hydraulic machines

Steam engines can be understood as the first modern machines used for the transformation of energy. Starting from the second half of the eighteenth century, they played a decisive role in the First Industrial Revolution and in the consequent rapid scientific and technological development that characterized the nineteenth century. Their use in the manufacturing processes of raw materials, consumer products and then in the field of transport and the production of electricity has had, as is well known, significant and irreversible effects which have determined the transition from a predominantly agricultural, artisanal and trade to the modern industrial system. In the Royal Application School for Engineers and Architects in Palermo, the first studies relating to steam engines date back to the mid-1800s, as evidenced by the oldest stationary machines and scientific equipment kept in the Museum (Fig. 1.1). The great Neville steam engine (Figs. 1.2 and 1.3) dates to that period, used, according to an industrial plant layout which remained practically unchanged until the first half of the 1900s, to activate the machinery in a production plant: the looms of a textile factory, the machine tools of a mechanical industry or pumps for extracting water in mines. Today, entirely restored, this great machine is equipped with an electric motor which allows it to rotate slowly to show the movement of the various organs, including its valve timing system. The collection also includes a rare "double expansion" steam engine with overlapping vertical cylinders, built by Willans & Robinson Limited (Fig. 1.4) in the second half of the nineteenth century. Due to a particular layout, these engines rotate faster than conventional models, representing an example of technical innovation improvement. In the same context, another find of considerable historical and technical interest is the large Ljungström steam turbine (Fig. 1.23) from the 1920s, coming from the ancient Alessandro Volta power plant in Palermo. This machine, very rare today, is characterized by a particular construction typology, which allows for a considerable reduction in size. Its historical relevance is instead due to the fact that it represents a piece of the city's history: it remained unharmed during the bombings of the Second World War, it supplied electricity to a large part of the city of Palermo until 1952. After the steam engines, the exhibition itinerary illustrates various examples of internal combustion engines, both gas and Diesel, which during the first half of the 1900s gradually replaced reciprocating (piston) steam engines in the naval and industrial sectors. Among these, the Langen & Wolf engine (Fig. 1.6) dating back to the last decade of the nineteenth century stands out for its rarity and value. Finally, the development of gas turbine engines for industrial use, which began at the end of the 1960s, is also represented in the collection by a rare example of the Mars turboshaft (Fig. 1.27), produced by Perkins Gas Turbine. The section dedicated to stationary engines also includes various hydraulic machines (1.1). Historically these were the first ones built for the transformation of energy, from that of water to mechanical energy. The Museum has specimens of the three most common types of hydraulic turbines (Francis, Kaplan, Pelton), used in the laboratory of the former Institute of Machinery for research and teaching activities, as early as the 1930s.

1.1 E.G. Neville & C.

1870s–1880s

The great "Neville" (Fig. 1.3) is a stationary steam engine, built by E.G. Neville & C. in Venice, in the second half of the nineteenth century. The engine has a Rider-type variable timing system, automatically controlled by a Buss-type centrifugal speed controller (Table 1.1). Starting from the Industrial Revolution, this type of machine was widely used, according to an industrial plant scheme which remained nearly unchanged until the first half of the 1900s, to drive various type of machinery, such as the looms of a textile factory, the machines tools of a mechanical industry, an electric generator or pumps for water extraction in mines. The engine is one of the older items in the Museum collection and, due to its awful and ruined conditions, it was fully restored in 2009.

Table 1.1 Technical specifications of the steam engine E.G. Neville & C.

Manufacturer	E.G. Neville & Co., Venice, Italy, second half of XIX century
Description	Stationary steam engine, single cylinder
Stroke	400 mm
Bore	204 mm
Displacement	26.1 L
Power	ca. 8 hp at 100 rpm with a steam pressure of ca. 8 bar
Valvetrain	Double slide valve with variable a Rider-type timing system, controlled by a Buss-type centrifugal speed regulator manufactured by Schaeffer & Budenberg (Fig. 1.2)
Lubrication	With multiple oilers (Fig. 1.2)

Fig. 1.2 Steam engine speed regulator

Fig. 1.3 The stationary steam engine by E.G. Neville & C., 1870s–1880s

1.2 Willans & Robinson Limited

1880s

Willans & Robinson Limited was an important and prolific UK company specialised in the construction of industrial and marine engines. Its production was characterized by the use of a particular innovative scheme, patented by the same company in 1884, in which the steam expansion is divided into two or three stages using coaxial cylinders (high, medium and low pressure respectively, see specifications in table 1.2). These sets of cylinders were aligned in groups of two or three on a common crankcase. Both inlet and discharge of the steam is governed by means of cylindrical distributors, coaxial to the pistons. Its particular and compact structure, in which the main moving parts are located inside the crankcase, allowed a better lubrication and, therefore, higher rotation speeds than other types of piston steam engines. The higher rotation speed made these engines suitable for the propulsion of small and medium-sized fast boats and to produce electricity, thanks to the direct and more efficiently driving connection to electric generator. For this reason, most of them have been produced for stationary use and are equipped with an automatic centrifugal speed regulator. Although this type of engine was produced in large quantities up to the 1920s, only a few examples have survived to this day. The original crankcase of the engine on display (Fig. 1.4) indicates its use in the industrial sector to produce electricity.

Table 1.2 Technical specifications of the steam engine Willans & Robinson Limited

Manufacturer	Willans & Robinson Limited, Thames Ditton, Surrey (UK), 1880s
Description	Stationary compound steam engine with two couple of in-line cylinders
Stroke	160 mm
Bore	181 mm (high pressure cylinders); 321 mm (low pressure cylinders)
Displacement	34.1 L
Power	13.6 hp at 450 rpm
Valvetrain	With two cylindrical distributors, coaxial to the pistons, directly driven by the crankshaft
Steam feeding	Supplied by steam at a maximum pressure of circa 10 bar, with an inlet valve controlled by a centrifugal speed governor driven by the crankshaft
Lubrication	Closed circuit inside the crankcase and splash-type lubrication

Fig. 1.4 The stationary steam engine Willans & Robinson Limited, 1880s of XIX century

1.3 Langen & Wolf

1880s–1890s

Starting from the end of the nineteenth century, the development of internal combustion engines led to the progressive and irreversible abandonment of reciprocating steam engines, both in the industrial sector and in means of transport. The stationary engine on display (Figs. 1.5 and 1.6) was built around 1890 by the Italian company Langen & Wolf of Milan (subsidiary of the Langen & Wolf Gasmotorenfabrik of Vienna). It represents one of the first models of internal combustion engines to be widespread and used as an alternative to steam engines. Its constructive scheme (Table 1.3) includes the main systems used in the first gas-fuelled engines. In particular, it is equipped with an archetypal valve train system with a "slider" intake valve, closely derived from those widely used in steam engines of the same period. The engine is equipped with a rudimentary ignition device, in which a small pilot flame (kept always lit) triggers combustion in each cycle when it comes into contact with the air-fuel mixture inside the cylinder, through the rapid and short opening of a small slider valve. This engine was originally used to operate a hydraulic pump in a private house.

Table 1.3 Technical specifications of the stationary gas engine Langen & Wolf

Manufacturer	Langen & Wolf, Fabbrica Italiana dei Motori a *Gaz* Otto (Fig. 1.5)
Description	Stationary four-stroke single-cylinder internal combustion engine
Stroke	180 mm
Bore	116 mm
Displacement	1.9 L
Power	1 hp at 230 rpm
Valvetrain	With an intake slide valve and an exhaust poppet valve
Fuel feeding	Supplied by gaseous fuel
Ignition	Fed by a pilot flame, controlled by a slide valve
Cooling	Liquid-fed
Lubrication	With multiple oilers

Fig. 1.5 The original plate of the Langen & Wolf engine

Fig. 1.6 The stationary gas engine Langen & Wolf, 1880s–1890s. Kind donation of Mrs. Gaetana Saporito Scalia

1.4 Ruston, Proctor & Co. CD

Early 1900s

Due to its construction scheme, the engine on display (Figs. 1.7 and 1.8) can be considered as the typical stationary internal combustion engine of the early twentieth century, widely used for multiple purposes in the industrial and civil fields. Although its overall structure is closely derived from modern steam engines, it has all those features that have been progressively developed over the following decades, including the spark ignition system (Table 1.4). This engine was built in England around 1906 by Ruston, Proctor and Co., one of the major engine builders at the time. It was originally used to operate the hydraulic pumps at a sulphur mine near Enna, in the middle of Sicily. Later, during the Second World War, it was transferred to a farm near Palermo, where it was used to produce electricity until the early 1950s.

Table 1.4 Technical specifications of the stationary gas engine Ruston, Proctor & Co. CD

Manufacturer	Ruston, Proctor and Co., Lincoln, United Kingdom, 1900s
Type	CD
Description	Stationary four-stroke single-cylinder internal combustion engine
Stroke	280 mm
Bore	181 mm
Displacement	7.2 L
Power	12 hp
Valvetrain	With poppet valves driven by a camshaft, with a centrifugal speed regulator (Fig. 1.9)
Fuel feeding	Supplied by gaseous fuel
Ignition	With a spark plug fed by an oscillating magneto
Cooling	Liquid-fed
Lubrication	With multiple oilers

Fig. 1.7 The original plate of the Ruston, Proctor & Co. engine

Fig. 1.8 The stationary gas engine Ruston, Proctor & Co. CD, early 1900s. Kind donation of Mrs. Gaetana Saporito Scalia

Fig. 1.9 Close-up views of the centrifugal speed regulator of the Ruston, Proctor & Co. engine

1.5 Heinrici Machinenfabrik-*40*

Early 1900s

Between the end of the 19th and the beginning of the twentieth century, the Louis Heinrici Machinenfabrik company produced a wide range of engines operating on the basis of the Stirling cycle (Fig. 1.10). The smallest ones, with a cylinder bore of 26 mm, developed a power output of a few fractions of a horsepower and were intended for operating little table-top equipment. The larger machines could be used to drive machinery tools in a production plant. The largest model had a cylinder diameter of 190 mm and developed a maximum power of half horsepower. Despite the poor efficiency, the constructive simplicity, the possibility of being fed with multiple low-grade fuels and their silent operation made Stirling engines particularly suitable for the use in factories or laboratories before the large-scale spread of electricity and related motors. The engine on display (Fig. 1.11) features a 40 mm cylinder bore (Table 1.5) and has been fully overhauled and tested in the museum's restoration laboratory.

Table 1.5 Technical specifications of the Stirling engine Heinrici Machinenfabrik-40

Manufacturer	Louis Heinrici Machinenfabrik, Zwickau, Germany, 1900s
Description	Stationary, single-cylinder, Stirling engine
Stroke	20 mm
Bore	40 mm
Displacement	25 cm^3
Power	0.02 hp at 350 rpm
Valvetrain	By sliding pistons
Fuel feeding	With a small burner
Cooling	Supplied by liquid or air
Lubrication	With multiple oilers
Mass	12 kg

Fig. 1.10 The Heinrici product catalogue of Stirling (hot air) stationary engines, 1914

Fig. 1.11 The stationary Stirling engine Heinrici Machinenfabrik-*40*, early 1900s. Kind donation of Mr Alessandro Pieri Vincentelli

1.6 Heinrici Machinenfabrik-*100*

Early 1900s

Founded in 1876, the Louis Heinrici Machinenfabrik company operated until the end of the 1920s, as one of the most important manufacturers of small and medium-sized stationary engines, operating on the basis of the Stirling cycle. Due to their constructive simplicity and a very silent operation, as well as the possibility to be fed with various fuels (both liquid and gaseous, even low-grades ones), these engines found a wide diffusion between the end of the nineteenth century and the beginning of the 1900s. Usually, depending on their size (classified according to the cylinder bore, in this case 100 mm), and so the power output supplied, they were used to drive industrial machinery, for small laboratory machinery (such as small mixers), as well as for domestic equipment such as water pumps or air cooling fans.

The Stirling engine, invented in 1816 by the Scotsman Robert Stirling (1790–1878), is based on a simple operating principle: the volume of gas contained within a closed cylinder is moved by a "displacer" piston alternately between a hot source and a cold capacity. The resulting change in gas pressure is used to drive an engine piston. Both pistons are connected to the crankshaft with of a 90° offset between the two cranks (Table 1.6). The engine on display (Fig. 1.12) comes from the historical collection of equipment of the Chemistry Laboratory founded by prof. Stanislao Cannizzaro at the University of Palermo in the second half of the nineteenth century.

Table 1.6 Technical specifications of the Stirling engine Heinrici Machinenfabrik-100

Manufacturer	Louis Heinrici Machinenfabrik, Zwickau, Germany, 1900s
Description	Stationary, single-cylinder, Stirling engine
Stroke	60 mm
Bore	102 mm
Displacement	490 cm^3
Power	0.125 hp at 350 rpm
Valvetrain	By sliding pistons
Fuel feeding	With a small burner
Cooling	Supplied by liquid or air
Lubrication	With multiple oilers
Mass	60 kg

Fig. 1.12 The stationary Stirling engine Heinrici Machinenfabrik-*100*, early 1900s

1.7 Hille Werke AG

1920s

The compression ignition engine, conceived and patented by Rudolf Diesel in 1892, found rapid diffusion in the industrial sector where, within a few years, thanks to its technological advantages, it replaced reciprocating steam engines. In the first half of the twentieth century, Diesel engines became gradually spread in the naval sector and later in the railway and automotive fields. The construction scheme of the first models derives strictly from that widely used for the other stationary engines, from which it differs mainly in the use of a fuel injection system, as well as in the shape of the combustion chamber (Table 1.7). The growing commercial demand led many mechanical companies to engage in the production and in the development of Diesel engines. Among these was Hille Werke AG in Dresden (Fig. 1.13), which produced the engine on display in the 1920s (Fig. 1.14). It was purchased to drive the hydraulic system in a farm. After its decommissioning it was donated to the University of Palermo and in 2010 it became part of the Museum's collection, where it was fully restored in between 2011 and 2013.

Table 1.7 Technical specifications of the Diesel engine Hille Werke AG

Manufacturer	Hille Werke AG, Dresden, Germany, 1920s
Description	Stationary, single-cylinder, 4-stroke Diesel engine
Stroke	300 mm
Bore	180 mm
Displacement	7.6 L
Valvetrain	With intake and exhaust poppet valves operated by pushrods and rocker arms
Fuel feeding	Supplied by a reciprocating single-cylinder pump for heavy oil (or Diesel fuel) direct injection, controlled by a Hartung-type centrifugal speed regulator (Fig. 1.15)
Cooling	Liquid-fed
Lubrication	With multiple oilers

Fig. 1.13 The original plate of the Hille Werke AG engine

Fig. 1.14 The stationary Diesel engine Hille Werke AG, 1920s

Fig. 1.15 Close views of the Hille Werke AG engine: cam and rocker arms for injection pump (on the left) and the Hartung-type centrifugal speed regulator (on the right)

1.8 Hanseatische Motorengesellschaft

1920s

Founded in 1916, the Hanseatische Motorengesellschaft (HMG) was one of the many German companies which in the first half of the twentieth century produced several Diesel engines for stationary plants and for motorboats. Initially the activity of HMG had been started for the construction of Callesen engines under license up to the end of the First World War, when it began the production of engines of its own design. A large part of HMG's production included "hot head" two-stroke Diesel engines, i.e. capable of operating even with poor quality fuels and high self-ignition temperatures. To this end, these engines, quite widespread in the first half of the 1900s, had a combustion chamber with a particular bulb shape (Fig. 1.17), which allowed, in cold start-up phases, an easily warm up by an external heat source. This heat source was then removed when the engine reached the thermal regime of normal operation, which was sufficient to keep the compression self-ignition process stable. In addition to their typical constructive robustness, stationary engines are characterized by the very frequent use of an automatic speed regulator (Table 1.8) which, in this case, controls engine operation by varying the quantity of fuel injected into the combustion chamber in each cycle.

Table 1.8 Technical specifications of the HMG Diesel engine

Manufacturer	Hanseatische Motorengesellschaft m.b.H., Hamburg-Bergedorf, Germany, 1920s
Description	Stationary single cylinder 2-stroke Diesel engine
Power	8 hp a 500 rpm
Air intake	Crankcase-scavenged type
Fuel feeding	Supplied by a reciprocating pump for heavy oil (or Diesel fuel) direct injection, controlled by a Hartung-type centrifugal speed regulator (Fig. 1.16)
Cooling	Liquid-fed by a single-cylinder reciprocating pump
Lubrication	Automatic with a Bosch mechanical oiler

Fig. 1.16 The fuel pump of the HMG engine controlled by the centrifugal speed regulator

Fig. 1.17 The stationary Diesel engine Hanseatische Motorengesellschaft, 1920s

1.9 Oreglia & C. 16 C

1920s

The mechanical workshop G. Oreglia & C. (Fig. 1.18) was one of the companies which, starting from the early 1900s, contributed to the mechanization of the agricultural sector in Italy. Founded in Piacenza by Giacomo Oreglia, the company remained active until the Second World War, producing stationary engines, tractors and agricultural machinery. Together with the Bubba company, in 1927 it was one of the first Italian manufacturers of tractors equipped with hot-head engines. The 16 C model (Fig. 1.19) is a rare survivor from this company's production. It is a four-stroke spark-ignition engine (Table 1.9), characterized by cylinders grouped in pairs in two independent blocks. These are fixed on a single base, in which are placed two camshafts. To allow the use of low-grade and cheaper fuels, the engine has a particular fuel system, in which the heat of the exhaust gases is used to improve the evaporation of the liquid fuel and therefore its mixing with the air. The engine is equipped with an automatic speed regulator, which controls the inlet throttle valve (Fig. 1.20). The large pulley connected to the flywheel, used to drive various machines by means of flat leather belts, as well as the massive flat-bottom engine support, are typical features designed for stationary applications. The engine on display comes from the Technical High School Vittorio Emanuele III in Palermo and underwent a complete restoration in 2012 at the museum laboratory.

Table 1.9 Technical specifications of the Oreglia & C. 16 C

Manufacturer	Officina Meccanica Ing. G. Oreglia & C., Piacenza, Italy, 1920s
Type	16 C
Description	Stationary, 4-inline-cylinder, 4-stroke, spark-ignition engine
Power	24 hp at 950 rpm
Valvetrain	With 2 side valves per cylinder, controlled by two camshafts housed in the crankcase, driven by gears
Fuel feeding	With a single barrel vertical evaporative-type carburettor for petrol or heavier liquid-type fuel
Ignition	With a spark plug per cylinder fed by a magneto
Cooling	Liquid-fed by a centrifugal pump
Lubrication	With a gear pump driven by the intake camshaft

Fig. 1.18 The original plate of the engine Oreglia & C. 16 C

Fig. 1.19 The stationary spark-ignition engine Oreglia & C. 16 C, 1920s

Fig. 1.20 The throttle valve at the air intake of the Oreglia & C. 16 C engine

1.10 DWK K.Z.22

1920s

The Deutsche Werke of Kiel was one of the largest state-owned industries in Germany, particularly active in the marine sector. Developed from the civil conversion of the Kaiserliche Torpedowerkstatt military shipyard at the end of the First World War, the company maintained a production strongly focused on the naval sector, expanding its production with a vast range of engines for marine and industrial use, from steam engines to large two-stroke Diesel engines. The K.Z.22 model (Table 1.10) belongs to this last category: built in the mid-1920s, the engine on display on display (Fig. 1.21 and Fig. 1.22) was used for research and educational activities at the Laboratory of engines of the University of Palermo until the end of the 1970s. For this reason, the engine is equipped with an ancient in-cylinder pressure gauge and together with a Ranzi hydraulic dynamometric brake to which it was coupled by means of a Pomini elastic coupling. Decommissioned in the 1980s, the engine underwent a conservative restoration at the museum laboratory in 2010.

Table 1.10 Technical specifications of the DWK K.Z.22

Manufacturer	Deutsche Werke Kiel AG, Kiel, Germany, 1927
Type	K.Z.22
Description	Stationary single cylinder 2-stroke Diesel engine
Power	16 hp a 525 rpm
Air intake	Crankcase-scavenged type
Fuel feeding	Supplied by a reciprocating single-cylinder pump for heavy oil (or Diesel fuel) direct injection, controlled by a centrifugal speed regulator
Cooling	Liquid-fed
Lubrication	Automatic with a Bosch mechanical oiler (Fig. 1.21)

Fig. 1.21 The Bosch automatic oiler lubricating up to six engine components

Fig. 1.22 The stationary two-stroke Diesel engine DWK K.Z.22, 1920s

1.11 STAL Ljungström Steam Turbine

1928

The turbine Ljungström owes its name to the Swedish engineers Birger and Fredrik Ljungström who patented it in 1908. These turbines were produced by the Svenska Turbinfabriks Aktienbolaget Ljungström (STAL) which they founded in 1913 and by other foreign companies under license, such as Ansaldo in Italy. Its uniqueness is that it is a multi-stage counter-rotary centrifugal steam turbine (Table 1.11). It consists of two discs of proper configuration facing each other, connected with two independent counter-rotating shafts. Crowns of blades are connected to the disks and the steam is introduced into the central compartment through hollow shafts, flowing towards the periphery. The counter-rotary turbines found large use in stationary powerplants, with some applications also in railway thanks to their favourable mass-to-power ratio. The limit to the number of radial stages, the need to adopt two separate electric machines and the growing demand for more powerful turbines determined the decline of the Ljungström turbines in favour of the traditional type from the end of the '50s. In 1928, the turbine on display on display (Fig. 1.23) was fitted in the former Alessandro Volta powerplant in Palermo (Sicily) where it remained in service up to 1952, feeding electricity to a large part of the city. The powerplant was equipped with two steam turbines: the Ljungström-STAL (9100 kW) and a Zoelly-type manufactured by Escher Wyss (5450 kW) from 1922.

Table 1.11 Technical specifications of the STAL Ljungström steam turbine

Manufacturer	S.T.A.L. (Svenska Turbinfabriks Aktienbolaget Ljungström), Stockholm and Finspång, Sweden, 1928
Type	Blading R 30 S 65-48/175 A. s/n 448
Description	Centrifugal counter-rotary steam turbine (Ljungström type), used in power plant for electricity production (typical layout shown in Fig. 1.24)
Feeding system	Supplied by superheated steam, with a maximum pressure of 14 bar
Cycle temperatures	Inlet steam temperature: 350 °C normal, max. 380 °C; condenser outlet temp.: normal 28 °C, max. 32 °C
Maximum power	9100 kW at 3000 rpm (maximum continuous power)
Control system	By means of one main supply valve and one steam control valve with a centrifugal governor
Lubrication	With a gear pump, feeding the main bearings of the turbine and the speed regulator
Condenser	One horizontal condensation chamber, positioned beneath the turbine, having a surface area of 800 m^2, fed by sea water supplied by a centrifugal pump (450 L/s) driven by an electric motor of 135 hp at 1470 rpm
Condensate extraction	With a centrifugal pump driven by an electric motor of 23.9 hp at 1440 rpm

Fig. 1.23 The STAL Ljungström steam turbine, 1928

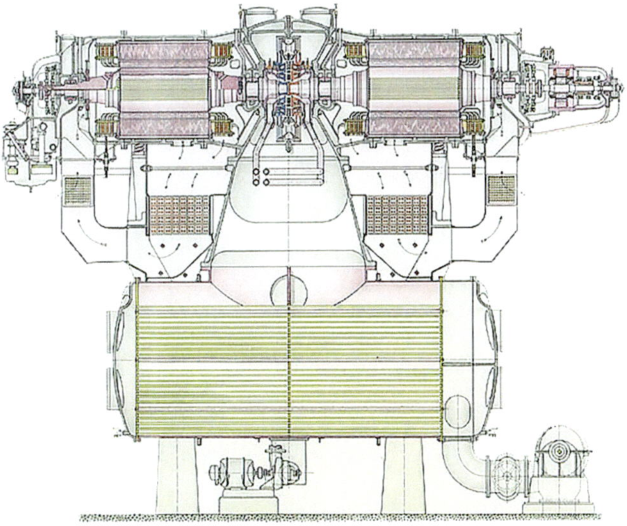

Fig. 1.24 Cross-section view of a Ljungström steam turbine with electric generators

1.12 Peroni & C. Duplex Steam Pump

First Half of XX Century

This particular machine (Fig. 1.25), named duplex pump, consists of a reciprocating steam engine, twin-cylinder with double-acting pistons, which drives a hydraulic pump, also twin-cylinder with double-acting pistons, mounted in tandem on the same frame (Table 1.12). Each piston of the engine is directly connected, by means of a steel rod, to the opposite piston of the pump. The stems also control the valve timing of the engine, through a series of connecting rods and rocker arms, while the pump is equipped with automatic valves for both charge and discharge. The pistons of the engine have a larger diameter than that of the pistons of the pump in order to ensure an operating pressure higher than that of the steam, allowing the feed of water into the steam generator where necessary. Compared to other systems, these pumps allowed a fine regulation of the feeding flow to the generator down to very low values, simply by acting on the inlet steam valve in the engine. On the other hand, they required a non-negligible steam consumption. Until the first half of the twentieth century these steam pumps were commonly used as a service or standby unit in industrial steam plants, in ships and in some locomotives.

Table 1.12 Technical specifications of the Peroni & C. duplex steam pump

Manufacturer	Peroni & C. s.p.a., Milan, Italy, first half of the XX century
Description	Reciprocating pump with two double-acting cylinders driven by a reciprocating steam engine with two double-acting cylinders (Fig. 1.26)
Steam cylinder bore	85 mm
Hydraulic cylinder bore	60 mm
Stroke	102
Steam feeding max. pressure	23 kg/cm^2
Pump feeding max pressure	27 kg/cm^2
Operating speed	From 40 to 80 per minute
Volumetric flow	1.04 litre/cycle

Fig. 1.25 The Peroni & C. duplex steam pump, first half of XX century

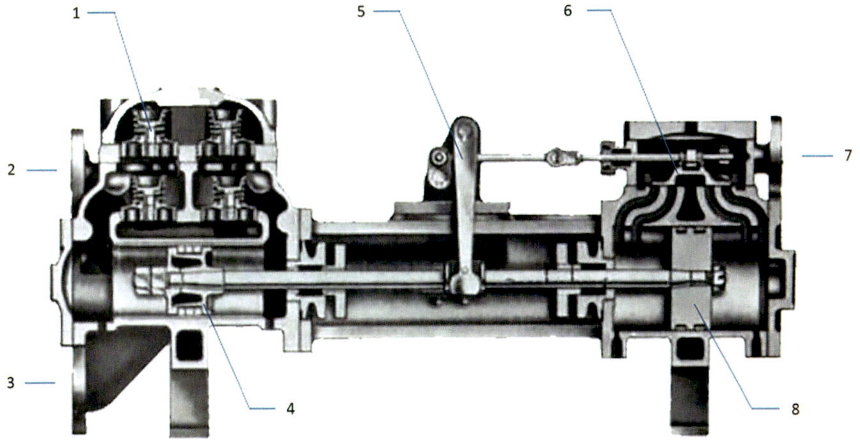

Fig. 1.26 Cross-section view of a duplex steam pump: 1) automatic charge and discharge valves of hydraulic pump; 2) water feeding flange; 3) water suction flange; 4) water pump piston; 5) driving rocker arm for engine slide valve; 6) slide valve which controls the steam flow in and out of the cylinder of the engine; 7) inlet steam flange; 8) engine piston

1.13 Perkins Gas Turbine Ltd. Mars

1963

At the end of the 1940s, after the initial development in aviation, gas turbine engines gradually found applications in other sectors, from the propulsion of vehicles to the drive of operating machines in both civil and industrial applications. For this reason, at the end of the 1950s, the English company Perkins Engines, already renowned for its Diesel engines, decided to create the Perkins Gas Turbine to undertake the production of the 50 hp Mars turboshaft, purchasing the construction license from the American company Solar Gas Turbines Limited of San Diego. The engine, strictly derived from the Solar T41 model, consists of a single-stage centrifugal compressor and a single-stage centripetal turbine, endowed with a simple bevel gears speed reducer (Table 1.13). Characterized by a very small mass and dimensions, the Mars engine was designed to drive hydraulic pumps, industrial compressors and electric generators. In particular, Perkins also attempted to commercialize it as the driving element of a fire-fighting hydraulic pump. Having lost the order from the British fire brigades due to its high operating noise, the engine remained in production until 1965, with a total of only 167 units. The Museum's Mars turboshaft (Fig. 1.27) was acquired in 1963 by the former Institute of Machinery of the University of Palermo for research and teaching activities.

Table 1.13 Technical specifications of the Perkins Mars turboshaft

Manufacturer	Perkins Gas Turbine Ltd., Peterborough, United Kingdom, 1963
Type	Mars
Description	Turboshaft with a single stage centrifugal compressor and a single stage centripetal turbine (Fig. 1.28)
Power	50 hp at 40,300 rpm
Fuel feeding	With a single injector fed with kerosene by means of a mechanical pump
Lubrication	With a gear pump
Transmission	With a bevel gear mechanism, with a reduction ratio of 10:1

Fig. 1.27 The turboshaft Perkins Gas Turbine Ltd. Mars, 1963

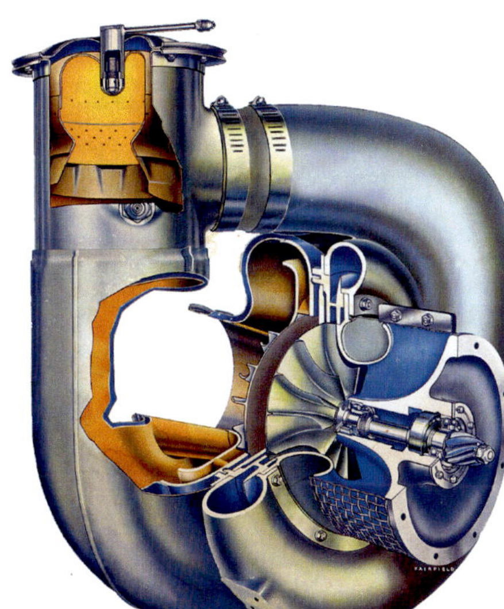

Fig. 1.28 Transparent view of the turboshaft Perkins Gas Turbine Ltd. Mars

1.14 San Giorgio Pelton Turbine

1938

The Pelton turbine, invented by Lester Allan Pelton in 1879, is a hydraulic machine still used nowadays to produce electricity. Given its operating characteristics, it is used in systems with water basins with high piezometric head (usually between 300 and 1400 m) and small flow rates (less than about 60 m^3/s); for this reason, it is quite widespread in mountain hydroelectric plants. There are various construction schemes, the most common one foresees an impeller with horizontal rotation axis and a single nozzle. However, there can be more than one nozzle: from a maximum of two in horizontal axis turbines, up to six in high power plants made with vertical axis turbines. The number of nozzles is limited by possible interference between the various jets, with the consequent reduction in efficiency. The regulation of the Pelton turbine usually takes place through the variation of the water flow made through the Doble needle. This consists of a shutter with a bulbous end placed in the centre of the nozzle. For a rapid reduction in the flow rate, a flow diverter is used, which intercepts, all or in part, the flow of water directed to the impeller blades. Manufactured in 1938 (Table 1.14), the turbine on display (Fig. 1.29) was used for research and then didactic activities at the Laboratory of Machinery of the University of Palermo.

Table 1.14 Technical specifications of the San Giorgio Pelton turbine

Manufacturer	San Giorgio, Rivarolo Ligure, Italy, 1938
Description	Pelton-type hydraulic turbine
Hydraulic head	15 m
Volumetric flow rate	7 L/s
Power	1 hp at 600 rpm

Fig. 1.29 The San Giorgio hydraulic Pelton turbine, 1938

1.15 Kaplan Turbine Rotor

1950s

The Kaplan turbine is a hydraulic machine invented by the Austro-Hungarian engineer Viktor Kaplan in 1913 and nowadays widely used to produce electricity from rivers or water basins characterized by small differences in piezometric head, up to ca. 10–70 m, but with great flow rates. The turbine on display (Fig. 1.30) has a 4-blade impeller with continuous variable pitch mechanisms for constant-speed load regulation as function of flow rate.

Fig. 1.30 The 4-blade variable pitch impeller for a Kaplan turbine, 1950s

1.16 Francis Turbine Rotor

1950s

The Francis turbine is a hydraulic machine invented by the British engineer James B. Francis in 1848 and nowadays widely used to produce electricity from rivers or water basins characterized by differences in piezometric head ranging from 10 to 700 m, with great flow rates. The turbine on display (Fig. 1.31) has a cast-iron impeller with 13 blades.

Fig. 1.31 The cast-iron impeller for a small Francis turbine, 1950s

Chapter 2
Marine Engines

Abstract The collection offers an overview into the evolution of marine engines, from steam engines to modern two-stroke diesel engines that are still used in large ships. In the XIX century the development of the steam engines revolutionized the naval sector by surpassing sailing navigation and leading to significant transformations in ship construction, transportation, and the global economy. The technical improvement of engines increased ship performance, such as speed, consumption, and range. Over time, piston steam engines were gradually replaced by steam turbines and diesel engines, up to the modern gas turbines derived from the aviation industry. Two-stroke diesel engines have found extensive use in the naval sector, with large engines exceeding 30 m in length and power output of ca. 4000 hp per each cylinder. These engines, particularly when coupled to turbomachinery and heat recovery systems, exhibit an overall efficiency greater than 50% and comparable to the modern stationary power plants.

The first core of the naval engine collection at the Museum was established starting in 2016, bringing together some artifacts that the University had possessed for several decades, based on their use in naval engines. Over time, additional engines donated by private individuals were gradually added to the collection. In 2020, the collection of naval engines expanded further with the acquisition of a group of ten highly historically and collectible significant artifacts, including the engine of the submarine "Medusa" from 1909. This expansion led to a comprehensive renewal of the related exhibition area. The growth of the collection was made possible through a collaborative agreement with the Heritage department of FIAT Chrysler Automobiles, now part of the Stellantis Group.

The most important pieces in this collection are those produced by Grandi Motori della FIAT, dating from the early 1900s to the late 1960s. They serve as a testimony to the extensive and prolific production of one of the largest naval engine manufacturers, not only in Italy but worldwide.

The naval engine section (Fig. 2.1) offers visitors the opportunity to observe the evolutionary journey of naval motorization, from steam engines to modern two-stroke Diesel engines still used in large ships. The practical application of

G. Genchi, *The Museum of Engines and Mechanisms of the University of Palermo*, History of Mechanism and Machine Science 48,
https://doi.org/10.1007/978-3-031-63042-2_2

Fig. 2.1 The exhibit area of the Museum dedicated to the marine machines

steam engines in the nineteenth century marked a pivotal turning point in the naval sector, as it replaced sail navigation and revolutionized ship construction, usage, and transportation, leading to significant changes in the global economy.

The performance development of motor ships in terms of speed, fuel consumption, autonomy, and more is closely connected to advancements in naval motorization. In less than a century, piston steam engines were surpassed by Diesel engines, and today, in some cases, modern gas turbines, sometimes adapted from the aerospace industry, are preferred. Among them, the two-stroke Diesel engine found its maximum application in the naval sector, with some of the largest engines of this type, reaching lengths of over 30 m and individual powers exceeding 4000 horsepower per cylinder. Their efficiency, especially when combined with turbo machinery and heat recovery systems for other onboard services, has reached values comparable to those of modern power plants.

2.1 Compound Steam Engine

Late XIX century

This steam engine, dating from the late nineteenth century, is the oldest exhibit in the Museum's collection of naval engines (Fig. 2.2). Although its exact origin is not known, the characteristics of various elements indicate a probable English construction. The engine is built according to a construction scheme widely used in the naval sector until the first decades of the 1900s, in which the cylinder block is connected to the crankcase by tie rods and interconnected supports. In this way the structure was much lighter and more accessible for maintenance than that typical of stationary engines, in which the cylinders and crankcase are cast in a single block. The engine has two cylinders with double-acting pistons and their diameter is different since the machine is of the compound1 type, with double steam expansion (Table 2.1). These characteristics, also widespread in the railway sector, allowed for a considerable reduction in overall dimensions for the same displacement and pressure of the fuel system. The engine is equipped with a Stephenson regulation system, which acts on

Table 2.1 Technical specifications of the compound steam engine

Manufacturer	Unknown, late XIX century
Description	Marine steam engine with two in-line double-acting cylinders and two stages of steam expansion
Stroke	160 mm
Bore	181 mm (high pressure cylinder), 321 mm (low pressure cylinder)
Displacement	34.1 L
Power	About 50 hp at 350 rpm with a steam pressure of 10 bar
Valvetrain	Two slider valves (high and low pressure), operated by a Stephenson mechanism, used as load regulator and speed reverser
Steam feeding	Supplied by steam at a maximum pressure of circa 10 bar
Lubrication	With multiple oilers

Fig. 2.2 Marine steam engine with two in-line double-acting cylinders and two stages of steam expansion. Manufacturer: unknown, late XIX century

the distribution timing to quickly and efficiently vary the power delivered and the direction of rotation of the machine, avoiding the use of a reverser for the propulsion propeller. Although in the past these engines were produced in large quantities, only a few examples of this size have survived to this day and are objects of collectible interest.

2.2 FIAT S 6185

1908

The S 6185 engine is one of the oldest produced by FIAT for the naval sector. The construction scheme adopted was quite modern for its time, with the use of various technical features (Table 2.2) capable of ensuring competitive performance and simplifying control and maintenance operations. The engine on display (Fig. 2.3) was produced by FIAT San Giorgio in 1908 for the submarine "Hvalen" of the Royal Swedish Navy (1909–1924), equipped with three independent propulsion units consisting of 6 S 6185 engines coupled in tandem. Each group was connected in series to an electric motor-generator and a propulsive propeller. The submarine was built by FIAT San Giorgio in the Muggiano shipyard near La Spezia; launched in 1909

Table 2.2 Technical specifications of the FIAT S 6185

Manufacturer	FIAT San Giorgio, Turin, 1908
Type	S 6185
Description	Four-stroke marine engine with six in-line cylinders coupled in tandem, cast-iron crankcase with single-piece cast-iron cylinder-head units, crankshaft with four main supports, riveted sheet metal oil pan (Fig. 2.4)
Stroke	200 mm
Bore	185 mm
Displacement	32.3 L
Power	150 hp at 750 rpm
Specific consumption	270 g/hp/h
Specific power	4.7 hp/L
Valvetrain	2 overhead valves per cylinder, operated by pushrods and rocker arms driven by an overhead camshaft
Fuel feeding	Single-barrel petrol carburettor Dimeler
Ignition	Single spark plug in each cylinder powered by a high-voltage magneto, with manual adjustment of the ignition timing
Cooling	Forced by gear pump with a seawater heat exchanger
Lubrication	Forced by mechanical pump with a cooling system
Starting system	By means of an electric motor-generator
Mass	1650 kg
Mass-to-power ratio	11.0 kg/hp

Fig. 2.3 The marine engine FIAT S 6185, 1908. Entrusted by Stellantis

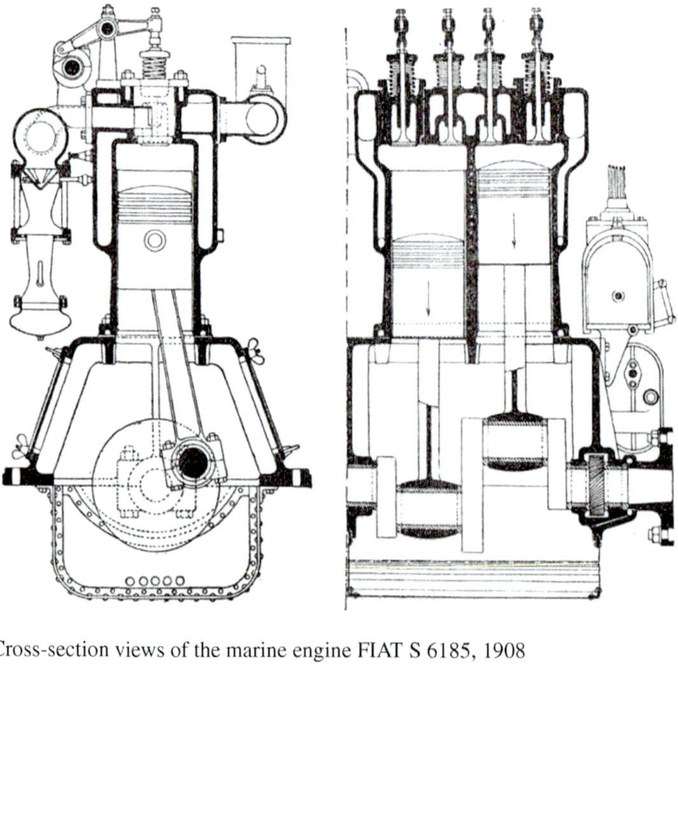

Fig. 2.4 Cross-section views of the marine engine FIAT S 6185, 1908

it remained in active service until 1919. Then it was used for other activities until it sank in 1924. In 1961 the wreck of the submarine was recovered, and FIAT took steps to acquire and restore the engine for display purposes, as evidence of its long activity in the naval sector.

2.3 FIAT 2 C. 116

1909

This engine is particularly important in the history of FIAT's Grandi Motori division as it represents the first model characterized by "modern" efficiency values (up to 260 g/hp/h) and it is one of the first FIAT 2-stroke engine (Table 2.3). The experiences gained in the study and construction of this engine have proved to be very useful for the development of the further model destined for both military and commercial applications, which allowed FIAT to reach a renowned position in the field of naval engines. Built by FIAT San Giorgio in 1909, the engine on display (Fig. 2.5) was used, paired with another identical unit, in the Medusa submarine of the Regia Marina (Fig. 2.6). In 1915 the submarine sunk during a war mission and its wreck was recovered in 1958, when FIAT purchased and restored the engine for display purposes, to show its fifty years long activity in the naval sector.

Table 2.3 Technical specifications of the FIAT 2 C. 116

Manufacturer	FIAT San Giorgio, Turin, 1909
Type	2 C. 116
Description	Two-stroke Diesel engine, with six in-line cylinders, cast-iron crankcase and cylinder blocks, crankshaft with eight main supports (including those of the air compressor)
Stroke	270 mm
Bore	250 mm
Displacement	79.5 L
Power	300 hp at 500 rpm (max. speed 550 rpm)
Specific power	3.8 hp/L
Air intake	Unidirectional inlet flow, forced by means of six alternative pumps (one for each cylinder) made up of the lower end of the engine pistons having an increased bore of 390 mm
Valvetrain	With variable timing system used both for engine load control and rotational direction change; 2 overhead intake valves in each cylinder, operated by an overhead camshaft; exhaust ports at the lower end of the cylinders, controlled by the piston stroke
Fuel feeding	Diesel fuel direct injection via pneumatic injectors, controlled by the camshaft;
Cooling	Forced by mechanical pump, with heat exchanger
Lubrication	Forced by mechanical pump with water-cooling circuit for the crankshaft main bearings
Starting	Pneumatic system
Mass	7 tons (including cooling and lubrication pumps)

Fig. 2.5 The marine engine FIAT 2 C. 116, 1909. Entrusted by Stellantis

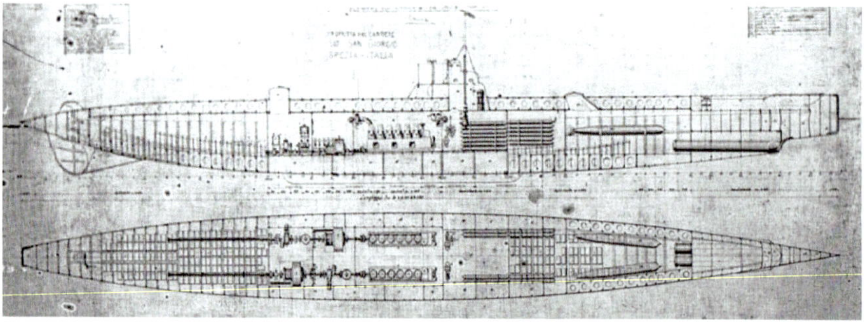

Fig. 2.6 Cross-section views of the Regia Marina submarine Medusa equipped with two FIAT 2 C. 116 engines

2.4 Archimede BS-2

1914–1931

At the beginning of XX century the use of two-stroke engines had spread rapidly also in the sector of small pleasure boats, thanks to their constructive simplicity (Fig. 2.7) and their ease of use and maintenance. These features (Table 2.4) made them particularly suitable for light and portable outboard motors. The Swedish Aktiebolaget Archimedes was one of the most renowned manufacturers of these type of engine thanks also to some successes in sporting competitions. In 1914 the Archimedes company proposed two models, the 2 hp BS and the 5 hp BS-2 for small pleasure

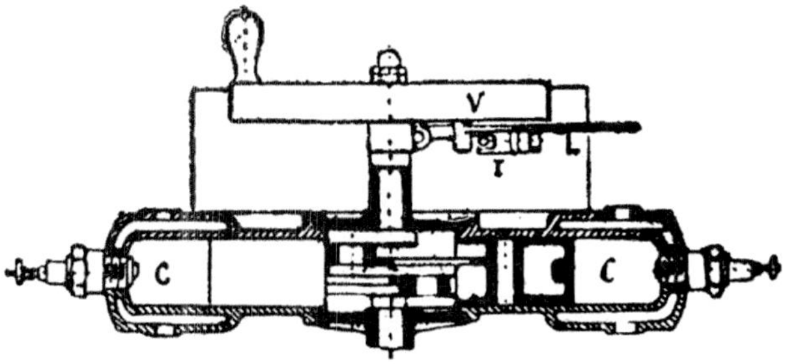

Fig. 2.7 Cross-section view of the outboard marine engine Archimede BS-2

Table 2.4 Technical specifications of the Archimede BS-2

Manufacturer	Società Industriale Italo-Svedese, Genoa, 1914–1931
Type	BS-2
Description	Two-stroke outboard marine engine with two opposed cylinders, cast iron crankcase and cylinders, flat-plane crankshaft (throws spaced 180° apart) with two main supports; three-bladed bronze propeller
Stroke	65 mm
Bore	70 mm
Displacement	500 cm^3
Power	5 HP at 1050 rpm
Air intake	Crankcase-scavenged type
Fuel feeding	Single-barrel petrol carburettor
Ignition	Single spark plug per cylinder powered by a high voltage distributor magneto, with manual adjustment of the ignition timing and engine rotational direction
Cooling	Seawater-fed, forced by a dynamic inlet scoop
Lubrication	By petrol and oil mixture (5%)
Mass	70 kg

boats up to 10 m in length. In Italy, these models were marketed by the Società Industriale Italo-Svedese in Genoa, as for the engine on display (Fig. 2.8).

Fig. 2.8 The outboard marine engine Archimede BS-2, 1914–1931. Kind donation of Mrs. Gaetana Saporito Scalia

2.5 Breda Roma

First half of the twentieth century

The Ernesto Breda Italian Company, founded in Milan in 1886, was one of the largest Italian companies in the metalworking sector, characterized by a vast and complex production of materials and machinery for heavy industry, shipbuilding, railway transport and aviation, with a wide range of engines, both of its own design and manufactured under licence. Although the information available to date does not allow the exact identification of the engine on display (Fig. 2.9), its technical characteristics (Table 2.5) and dimensions suggest that it was designed for the use in small boats, such as fishing boats and tugboats. The engine was later used as a stationary machine for didactic purpose at technical high school in Palermo. The monogram of the manufacturer's name with the word *Roma* are shown on the base covers, probably referring to the production site or a specific application.

Fig. 2.9 The marine engine *Roma* manufactured by Breda Company, between 1910 and 1920s

Table 2.5 Technical specifications of the Breda *Roma*

Manufacturer	Società Italiana Ernesto Breda per Costruzioni Meccaniche, between 1910 and 1920s
Description	Inline 4-cylinder 4-stroke Diesel engine, crankcase, cylinders, and head made of cast iron, crankshaft with five main bearings
Valvetrain	2 overhead valves per cylinder, operated by pushrods and rocker arms driven by a camshaft housed in the crankcase
Fuel feeding	Direct injection system, supplied by a 4-piston reciprocating fuel pump
Cooling	Liquid-fed
Lubrication	Forced by a gear pump

2.6 Niagara Motors Special

1920s

At the beginning of the twentieth century, the rapid technical progress of internal combustion engines, especially four-stroke petrol engines, gave a notable boost to the development of recreational powerboating. The first models, often derived from automotive engines, were soon followed by new engines specifically designed for small and medium-sized motorboats. In Europe and North America various companies began to specialise in marine engines, promoting their product by sporting activities. One of the most active among these was Niagara Motors, with a vast range of models,

Fig. 2.10 The marine engine Niagara Motors Co. Special, 1920s

Table 2.6 Technical specifications of the Niagara Motors Special

Manufacturer	Niagara Motors Corp., Dunkirk, N.Y. (USA), 1920s
Type	Special
Description	Inline 4-cylinder 4-stroke engine, crankcase, cylinder block and head made of cast iron
Stroke	100 mm
Bore	70 mm
Displacement	1539 cm^3
Power	15 hp
Valvetrain	2 side valves per cylinder, operated by single camshaft
Fuel feeding	Zenith single-barrel petrol carburettor
Ignition	Single spark plug per cylinder powered by a high-voltage magneto
Cooling	Liquid-fed
Lubrication	Forced by a gear pump

from 2 to 8 cylinders, with a range of power output between 5 and 200 hp. The engine on display (Fig. 2.10) is a 15 hp model, named Special (Table 2.6).

2.7 FIAT V 123

1929–1930s

In the second half of the 1920s FIAT began the technical development and then the production of a series of medium and high speed four-stroke Diesel engines, equipped with direct injection system fed by a mechanical piston pump. The characteristics and performance of these engines made them particularly suitable for both industrial and naval use. Their construction was based on a semi-modular layout which allowed the production of a vast range of engines with different displacement and maximum power output ranging from 24 to 140 hp. This was achieved through the alternative use of 3 different cylinder bore (120, 140 and 170 mm) and a variable number of in-line cylinders from 2 to 6. The engine on display (Fig. 2.11), named V 123, has 3 cylinders with 120 mm bore (Table 2.7). Its design allowed the use as marine or stationary engine, without significant modifications.

Table 2.7 Technical specifications of the FIAT V 123

Manufacturer	FIAT, Grandi Motori factory, Turin, from 1929 up to the 1930s
Type	V 123
Description	4-stroke engine with 3 in-line cylinders, crankcase, cylinders, and heads made of cast iron, crankshaft with four main bearings (Fig. 2.12)
Stroke	180 mm
Bore	120 mm
Displacement	6.1 L
Power	30 hp at 800 rpm, max. 36 hp at 1000 rpm
Specific power	5.9 hp/L
Valvetrain	2 overhead valves in each cylinder, controlled by rods and rocker arms with camshaft in the crankcase driven by a gears and chain transmission
Fuel feeding	Direct in-cylinder injection fed by a Diesel–fuel pump with 3 pistons (one pump for each cylinder), controlled by a centrifugal speed regulator
Cooling	Liquid-fed forced by mechanical pump
Lubrication	Forced by a gear pump
Starting	Pneumatic, direct in-cylinder injection of compressed air
Mass	1100 kg (including flywheel)
Mass-to-power ratio	30.6 kg/hp

Fig. 2.11 The marine engine FIAT V 123, 1929–1930s. Entrusted by Stellantis

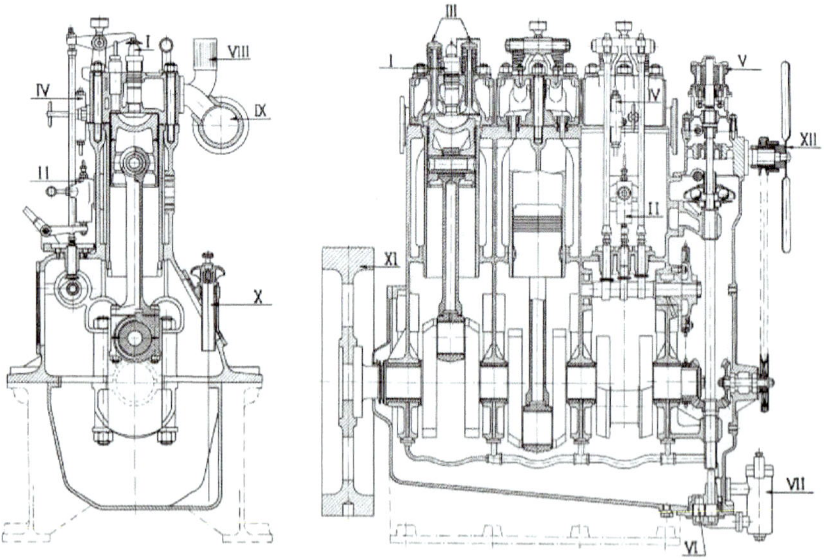

Fig. 2.12 Cross-section views of the engine FIAT V 123

2.8 FIAT V 1612

1936–1940s

In 1932 FIAT began the development of a new series of "high-speed", four-stroke Diesel engines with direct injection, characterized by a 60° V-shaped arrangement of cylinders with a bore of 160 mm and a stroke of 180 mm (Table 2.8). These engines were produced in large numbers in two versions with 12 cylinders (V 1612) and 16 cylinders (V 1616). Thanks to their mass-to-power ratio, engines of both types were used in the naval sector, for various fast motorboats, patrol boats and minesweepers, as well as in the railway sector. The ATR.100, a 3-body articulated railcars, were equipped with two V1612 engines allowing a maximum speed of over 160 km/h and a cruising speed limited to 120 km/h. From 1935 to 1961, the ATR.100 s remained in active service, with the same type of engines. After World War II, the V 1612 s were also used to repower 12 heavy shunter locomotives Whitcomb 65DE14. The engine on display (Fig. 2.13) is a railway type V 1612 and its crankshaft (Fig. 2.14) is shown separately for educational purposes, together with its connecting rods and pistons.

Table 2.8 Technical specifications of the FIAT V 1612

Manufacturer	FIAT, Grandi Motori factory, Turin, from 1936 up to the end of the 1940s
Type	V 1612
Description	4-stroke V-60° 12-cylinder engine, crankcase and oil sump made of welded steel sheets, crankshaft with seven main bearings, aluminium alloy pistons
Stroke	180 mm for pistons with main connecting rods (left-side bank), 186 mm for pistons with secondary connecting rods (right-side bank)
Bore	160 mm
Displacement	44.2 L
Power	350 hp at 1400 rpm, max. 540 hp at 1800 rpm
Specific power	Max. 12.2 hp/L
Valvetrain	4 overhead valves per cylinder, controlled, via pushrods and rocker arms by a single camshaft between the two banks
Fuel feeding	Diesel fuel direct injection fed by a 12-piston pumps (one for each cylinder) with a maximum pressure of ca. 230 bar, controlled by a centrifugal speed regulator
Cooling	Liquid-fed, forced by a mechanical pump
Lubrication system	Forced by a gear pump
Mass	2500 kg
Mass-to-power ratio	4.63 kg/hp

Fig. 2.13 The marine and railway engine FIAT V1612, 1936—the end of the 1940s. Entrusted by Stellantis

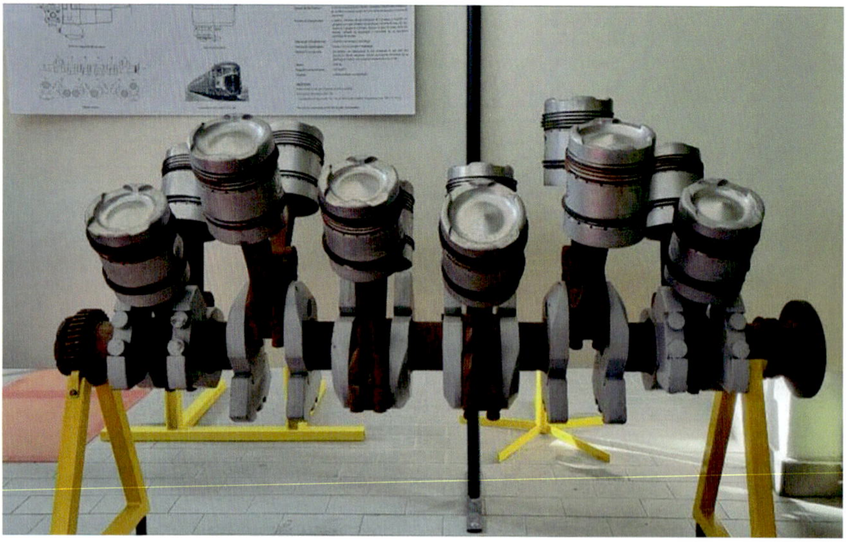

Fig. 2.14 The crankshaft of the FIAT V1612 engine with connecting rods and pistons. Entrusted by Stellantis

2.9 Scale Model of the Cargo Ship Italterra

1940s–1950s

The Italterra was a cargo ship used up to the 1950s to transport FIAT cars to the United States. Originally the ship was part of the vast fleet of Liberty-class vessels (Fig. 2.15) built in the United States and Canada in the first half of the 1940s. During World War II they were used to transport materials and armaments across the Atlantic, for supporting the Allied activities in Europe. At the end of the conflict some ships were sold to Italy in order to support the recovery of economic activities. At the end of the 1940s one of them, renamed Italterra (Fig. 2.16), was modernized and modified to allow the transport of 1000 cars. Furthermore, to make its operational use more efficient and economical, in 1953 the original triple expansion steam engine, delivering 2500 hp at 76 rpm, was replaced with a modern in-line six-cylinder two-stroke Diesel engine produced by FIAT. The new engine, model A 686, had a normal service power of 3600 hp at 125 rpm, with a significantly reduced specific consumption. Thanks to this engine, the ship could be repowered in just 45 days without any major changes to the rest of the ship. Fuel consumption and operating costs have been reduced and cruising speed has increased from 10 to approximately 12 knots. The scale model on display (Fig. 2.15) represents the central part of the ship, with particular regards for the engine room (Table 2.9) before the upgrade with the FIAT Diesel engine. The model was made by the Giovanni Feré company of Cinisello Balsamo (Milan) upon a request of FIAT, which used it for promotional purposes in international exhibitions and congresses, including the XXII Salon Nautique International (Boat Show) held in Paris in 1956.

Table 2.9 Technical specifications of the scale model of the cargo ship Italterra

Scale model builder	Giovanni Feré company of Cinisello Balsamo, Milan, 1954
Description	Scale model of the central part of a Liberty-class cargo ship, partially sectioned to show the engine room
Propulsion	Triple expansion steam engine, with operating pressure 15 bar, 2500 hp at 76 rpm, specific consumption 600 g/hp/h, driving a single propeller for a cruise speed of 10–11 knots
Max. length	134 m
Max. width	17.4 m
Displacement	Approximately 12,950 tons (full load)

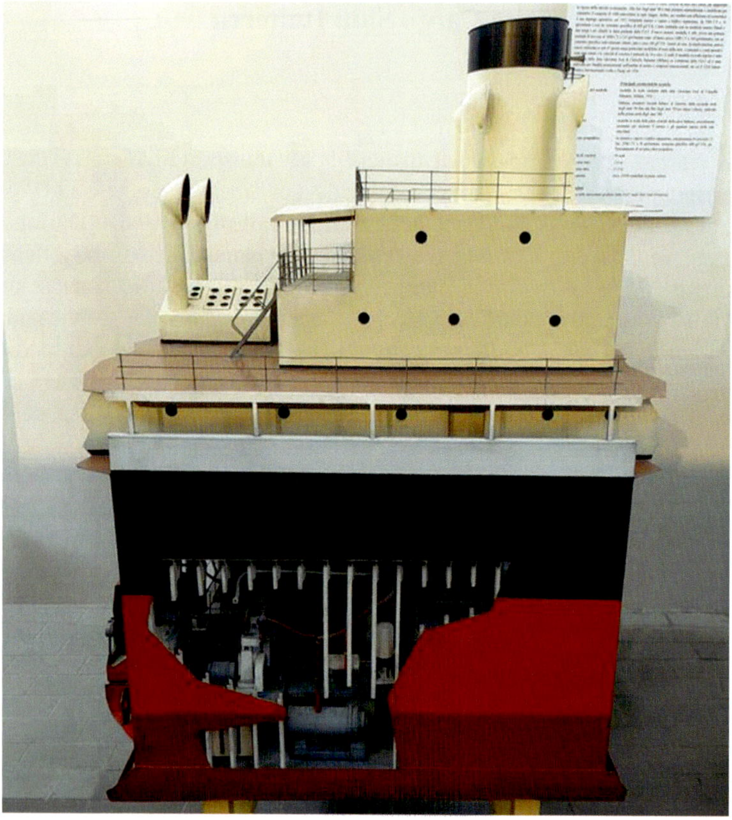

Fig. 2.15 The scale model of the central part of the cargo ship Italterra, partially sectioned to show the engine room. Entrusted by Stellantis

Fig. 2.16 The cargo ship Italterra

2.10 FIAT 480 Engine Piston

1952–1960s

The marine engines of the 480 series were produced by FIAT's Stabilimento Grandi Motori starting from 1952. They were characterized by the relatively small value of the piston stroke compared to the diameter of the cylinders. This design, defined as "short stroke", was chosen to make the use of these engines particularly suitable for ferries and passenger ships, in which the configuration of the various decks is such as to maximize the space on board for increasing the payload. This often implies the reduction of the height of the machine room and therefore requires the use of engines with reduced dimensions and masses. Thanks to these characteristics, the 480 series engines had a good commercial spread in the 1950s and 1960s. These engines could be produced with a variable number of cylinders according to operational requirements and could provide a power output of 300 hp per cylinder. Subsequently, various supercharged models were developed, featuring a power output of 600 hp per cylinder. The normal operating speed of 270–280 rpm allowed both a direct coupling transmission to the propeller and the use of a speed reducer. The piston on display (Fig. 2.17) has been partially sectioned by FIAT for educational purposes, allowing it to show its structure designed for crosshead joint commonly used as part of the slider-crank linkages in long reciprocating engines, and internal oil passages used both for lubrication and cooling (Table 2.10).

Table 2.10 Technical specifications of the FIAT 480

Manufacturer	FIAT, Grandi Motori factory, Turin, from 1952 until the end of the 1960s
Description	Piston of a two-stroke marine Diesel engine, main structure made of steel with lower skirt made of aluminium alloy, overall height 1885 mm
Stroke	640 mm
Bore	480 mm
Displacement	115.8 L per cylinder
Power	300 hp/cylinder at 280 rpm (aspirated version), 600 HP/cylinder at 280 rpm (more powerful supercharged version)
Specific power	2.59 hp/L, 5.18 hp/L (supercharged version)

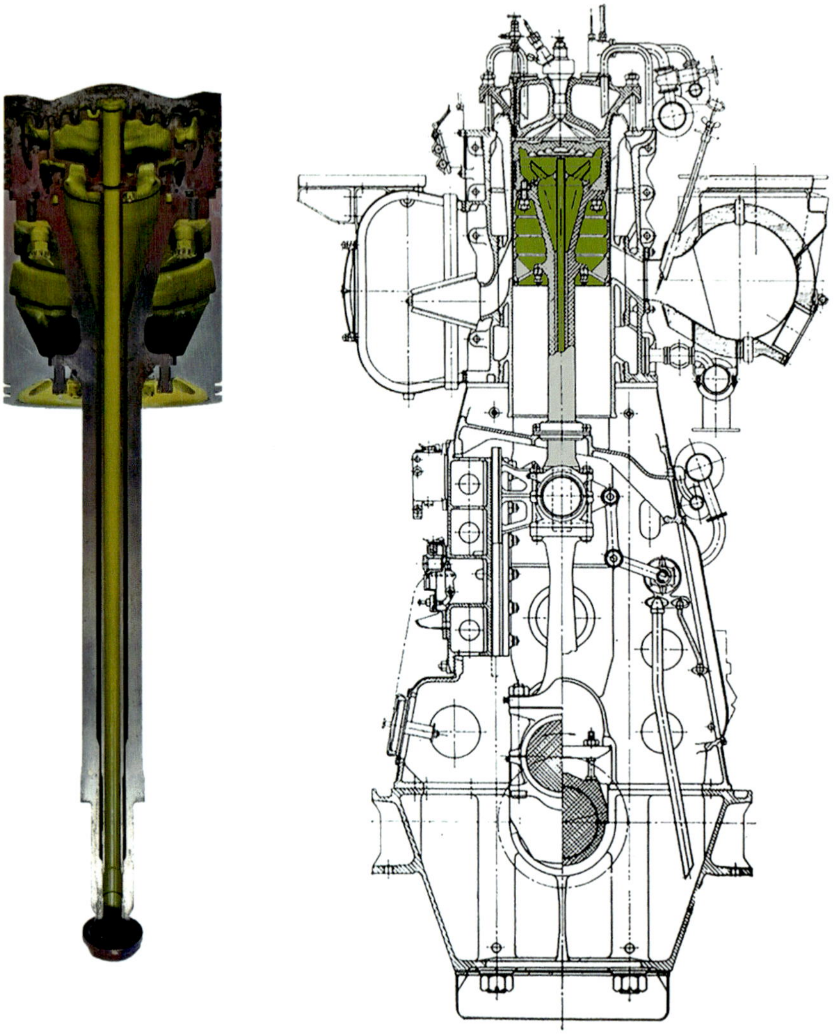

Fig. 2.17 The piston of a marine engine FIAT 480 (on the left) compared with the cross-section view of the entire engine (on the right). Entrusted by Stellantis

2.11 Burmeister & Wain Engine Exhaust Valve

1950s

Burmeister & Wain was a well-known Danish company, already active in the naval sector since the mid-nineteenth century. At the beginning of the twentieth century, it was among the first to give a strong impulse to the development and diffusion of Diesel engines, for industrial and marine applications, making a notable contribution to the technological progress of two- and four-stroke Diesel engines (Table 2.11 and Fig. 2.18).

Table 2.11 Technical specifications of the exhaust valve of a Burmeister & Wain engine

Manufacturer	Burmeister & Wain, Denmark, 1950s
Description	Exhaust valve (Fig. 2.19) for a uniflow-scavenging-type combustion chamber of a large marine two-stroke Diesel engine

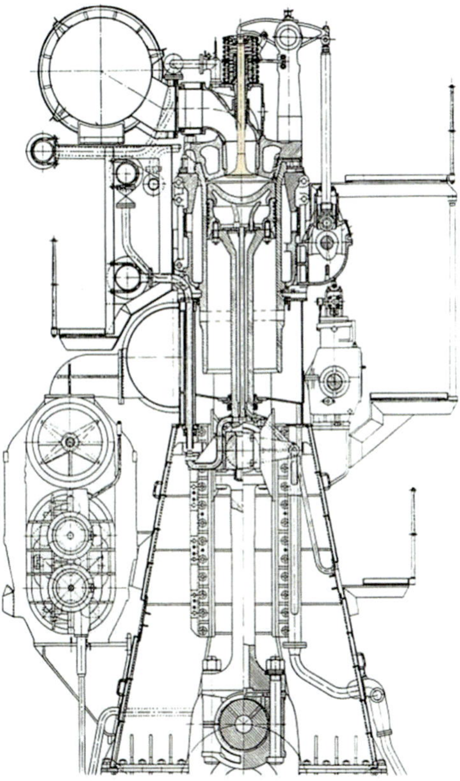

Fig. 2.18 Cross-section view of a two-stroke Burmeister & Wain marine engine, showing the uniflow-scavenging combustion chamber with the over-head exhaust valve

Fig. 2.19 Exhaust valve for a uniflow-scavenging-type combustion chamber of a large marine two-stroke Diesel engine. Burmeister & Wain, Denmark, 1950s. Kind donation of prof. Augusto Ajovalasit

2.12 British Seagull Silver Century

1960s

The Silver Century (Fig. 2.20) is a small single-cylinder outboard engine, produced from the mid-1960s and belonging to the vast series of two-stroke marine engines built for over fifty years by British Seagull. Characterized by great construction simplicity (Table 2.12) and renowned reliability, these engines have been widely used, since the 1930s, both as main engines and as auxiliary engines, in various types of pleasure boats. Due to their characteristics, nowadays many engines are still used by enthusiasts and collectors all over the world.

Table 2.12 Technical specifications of the British Seagull Silver Century

Manufacturer	British Seagull Co., Poole (Dorset), United Kingdom, 1960s
Type	Silver Century
Description	Two-stroke outboard marine engine with one horizontal cast-iron cylinder, crankcase and cylinder made of aluminium alloy, five-bladed bronze propeller
Stroke	57 mm
Bore	40 mm
Displacement	102 cm^3
Power	4.5 hp at 4500 rpm
Specific power	44.1 hp/L
Air intake	Crankcase-scavenged type
Fuel feeding	Single-barrel petrol carburettor
Ignition	One spark plug per cylinder powered by a high voltage distributor magneto
Cooling	Liquid-fed by a centrifugal pump
Lubrication	Supplied by a premixed petrol-oil mixture (1:10 in volume)
Starting	Manual pull start, via rope wound on the flywheel
Mass	17 kg (dry)
Mass-to-power ratio	3.78 kg/hp
Propeller transmission ratio	12/48 by means of a reduction drive with a pair of bevel gears

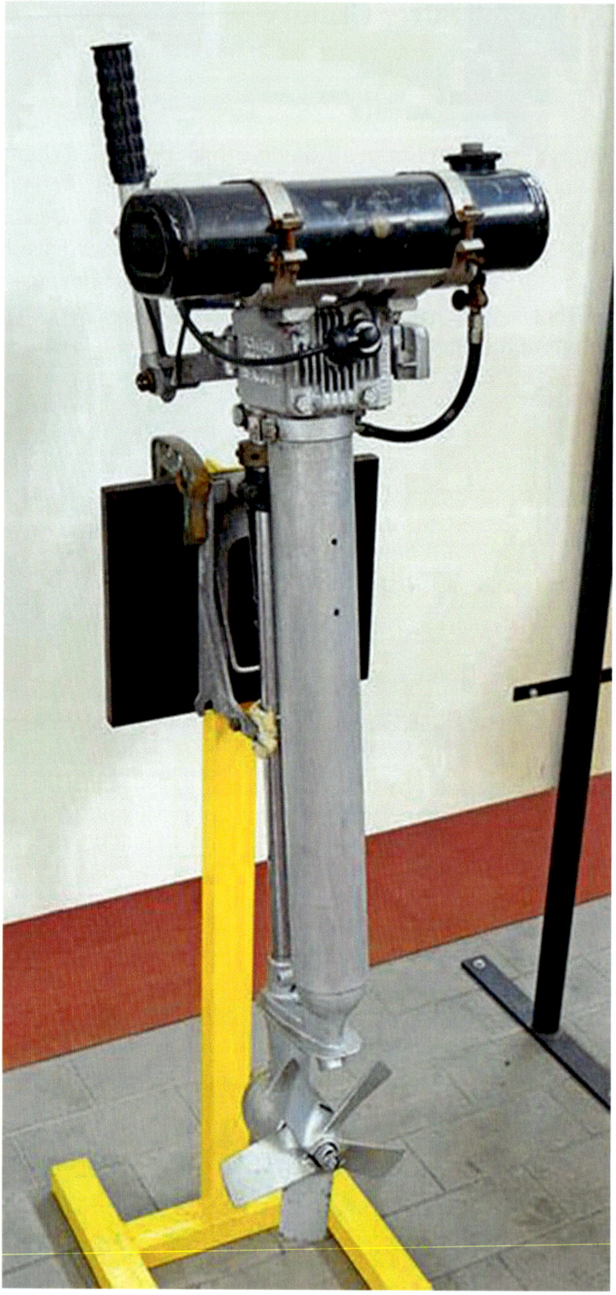

Fig. 2.20 The outboard marine engine British Seagull Silver Century, 1960s. Kind donation of Mr. Giosuè Lo Piccolo

2.13 Scale Model of the FIAT 1060 S Engine

1968

The 1060 S was the largest and most powerful marine engine built by FIAT Grandi Motori, with a max. Power output of 3800 hp per cylinder. Designed in the second half of the 1960s, it represents the pinnacle of the technical development of large two-stroke supercharged Diesel engines carried out by FIAT. Its name come from the cylinder bore dimension of 1060 mm. The research and development prototype, built with two-inline cylinders, was built and proved at the test bench in Turin, in 1968. Then, the production of complete engines for operational use on large cargo and oil tanker ships was carried out, in small series, in Grandi Motori plant in Trieste, where, in the early 1970s, FIAT had transferred most of its business in the naval engine sector. None of those engines produced has survived and the scale model on display, built in the late 1960s for promotional activities, accurately represents the twin-cylinder prototype. As well shown by the scale model (Fig. 2.21), the 1060 S was made by means of a modular arrangement (Table 2.13) giving the possibility of realising up to 12-cylinder engine (Fig. 2.22), as function of the power output required.

Table 2.13 Technical specifications of the FIAT 1060 S

Manufacturer	FIAT, Grandi Motori Plant, Turin, 1968
Type	1060 S
Description	Two-stroke Diesel marine engine with two in-line cylinders, engine frame made of welded sheets pre-stressed by steel tie rods, composite crankshaft
Stroke	1900 mm
Bore	1060 mm
Displacement	1677 L per cylinder
Power	3800 hp per cylinder at 106 rpm
Specific consumption	Circa 150 g/hp/h
Specific power	2.27 hp/L
Air intake	Crossflow scavenging arrangement with intake and exhaust ports located at the base of the cylinders, inlet air washing flow forced by supercharging system
Supercharging	Two intercooled stages in series, the first one with turbocharger powered by exhaust gases, the second one by reciprocating washing pumps (one per cylinder, driven by the engine piston)
Fuel feeding	Direct injection with a pneumatic nozzle per cylinder
Cooling	Forced by pumps with seawater heat exchangers, additional piston cooling by lubricating oil
Lubrication	Forced by pumps with heat exchanger

Fig. 2.21 The scaled model of the 1968 twin-cylinder prototype of marine engine FIAT 1060 S. Entrusted by Stellantis

Fig. 2.22 A 12-cylinder marine engine FIAT 1060 S during the assembling process. Actual dimensions: width 8.49 m, height 12.85 m, length 24.92 m. Engine mass 1640 tons

Chapter 3
Aircraft Engines

Abstract The Museum houses a significant collection of aircraft engines, showcasing the technical evolution in aviation history from the pioneering days after the Wright brothers' first flight up to the modern turbojet engines. The most used arrangements and solutions over different periods are well represented, ranging from air-cooled radial engines to liquid-cooled in-line and V-shaped cylinders engines, including rotary engines and experimental prototypes. Recent restoration activities, carried out as part of a wide cooperation project with the General Staff of the Italian Air Force, further enhanced the museum exhibit, with rare engines and prototypes from the early 1900s to modern jet engines.

G. Genchi, *The Museum of Engines and Mechanisms of the University of Palermo*, History of Mechanism and Machine Science 48,
https://doi.org/10.1007/978-3-031-63042-2_3

Fig. 3.1 The exhibit area of the Museum dedicated to the aircraft engines

After the Wright brothers' first flight in 1909, the aeronautical industry performed a rapid development which the first piston engines, often of automotive derivation, underwent enormous technical and performance evolution until the advent of turbojet engines, since the end of the Second World War. Particularly ingenious construction solutions date back to the first pioneering period of aeronautics, including that of rotary radial engines, in which the propeller rotated together with the cylinder block in order to improve engine cooling. To this category belongs the French engine Le Rhône Jby of 1917 displayed at the Museum (Fig. 3.1) as well as another of its rarest and most prestigious specimen, the Siemens Halske Sh.IIIa counter-rotary engine of 1918, which can be considered as the latest evolution of rotary radial engines. Other important exhibits in the collection come from Germany, including some Daimler Mercedes engines, which are part of a vast set of technical material given to Italy for research activities, as part of war reparations for the First World War. Their arrival in Palermo is linked to that of Antonio Capetti, professor in charge of Aircraft Engines at the Polytechnic of Turin (whose later he was Rector), who in 1925 was appointed professor at the chair of Thermal and Hydraulic Machines at the Royal University of Palermo. For each historical period, the collection includes at least one representative specimen of the main technical solutions adopted and in the various intended uses. From air-cooled engines, with the classic radial arrangement of the cylinders, and liquid-cooled engines with in-line and V-shaped cylinders. Over time, both solutions have been alternatively adopted by the main manufacturers, for engines intended for light aircraft and trainers up to large engines used in World War II. Some of important specimens in the collection belong to the latter period, including an original Daimler Benz DB605, technically advanced in its time, used in various German and Italian fighter planes such as the Messerschmitt Bf 109, the FIAT G 0.55, the Reggiane 2005 and the Macchi 205. The vast production of FIAT in the aeronautical field is well represented by some engines that are quite rare today including a FIAT A 74 and a FIAT A 80. Finally, the following Cold War period belongs to the American jet engine General Electric J47. Like others, the latter is exhibited with some parts partially disassembled for educational purposes. Since 2021, the exhibit layout of the aeronautical section has been considerably enriched thanks to the introduction of numerous unique or very rare specimens, which the General Staff of the Italian Air Force has entrusted to the Museum as part of a vast and articulated cooperation project between two entities, aimed at research, protection, and enhancement of the historical heritage. Numerous engines, coming from the deposits, have been restored for exhibit purpose. Ranging from radial engines from the beginning of the 1900s to modern jet engines, this collection includes experimental prototypes, as well as some of the most important models produced by the main Italian manufacturers, including FIAT, Alfa Romeo, Piaggio and Isotta Fraschini, and foreigners such as Mercedes, Rolls-Royce and General Electric.

3.1 Anzani A2

1909

Alessandro Anzani was one of the most important pioneers in the development of aero engines. Born in Milan (Italy), he moved to France in the early 1900s, where he initially put efforts into various motorsport activities, including motorcycling, by means of the construction of engines of own original design. Soon he entered the rising aeronautical sector by starting the development and the series production of a wide and renowned range of engines, both for military and general aviation aircraft. Many of them were used in various competitions and aviation feats, including the first aerial crossing of the English Channel, carried out in 1909 by Luois Blériot with an own aircraft (Blériot XI) equipped with an Anzani engine. The first models, derived from motorcycle engines, had 3 cylinders arranged in a double V of 60°. Then Anzani developed the new model A2 (Fig. 3.2), based on an inverted Y layout with 3 cylinders spaced of 120° (Table 3.1). All were air-cooled with fixed cylinders, unlike other manufacturers who preferred the rotary engine arrangement. To obtain further power improvement, Anzani also developed engines with a greater number of cylinders, characterized by innovative technical features.

Table 3.1 Technical specifications of the Anzani A2 engine

Manufacturer	Anzani Moteurs d'Aviation, Asnières (later Courbevoie), Paris, 1909
Type	A2
Description	Aircraft engine with 3 radial cylinders spaced of 120°, aluminium alloy crankcase, each cylinder is made of cast-iron together with its head
Stroke	120 mm
Bore	105 mm
Displacement	3117 cm^3
Power	30 hp at 1300 rpm
Specific power	9.6 hp/litre
Valvetrain	2 overhead valves in each cylinder; automatic intake valves and exhaust valves controlled via pushrods and rocker arms by a bud wheel housed in the crankcase
Fuel feeding	Single-barrel petrol carburettor
Ignition	One spark plug per cylinder with a high voltage magneto
Cooling	Air-fed
Lubrication	Forced by means of a gear pump
Mass	48 kg
Mass-to-power ratio	1.6 kg/hp
Application	Blériot, Deperdussin, and Sommer aircraft

Fig. 3.2 The aircraft engine Anzani A2, 1909. Entrusted by Aeronautica Militare

3.2 Mercedes D.IV

1915–1916

The Mercedes D.IV is an 8-inline-cylinder aircraft engine (Figs. 3.3 and 3.4) manufactured by the German company Daimler Motoren Gesellschaft during the First World War. The D.IV was the first mass-produced German engine equipped with a propeller reduction drive. Its construction scheme derived from the previous and very successful D.III, model from which it differed for the larger number of cylinders (from 6 to 8). The engine succeeded the clearance test in December of 1915, but it was produced in only 429 copies because of its poor reliability, mainly due to the fragility of the crankshaft for its excessive length. Because of this the D.IV was soon replaced by a completely redesigned engine, the D.IVa with 6 cylinders. The designation similar between the two engines is due to the classification then in use in Germany, based on increasing level of power output (by I to VI), in which the first letter indicates the manufacturer (Table 3.2).

Table 3.2 Technical specifications of the Mercedes D.IV

Manufacturer	Daimler Motoren Gesellschaft, Stuttgart-Untertürkheim, Germany, 1915
Type	Mercedes D.IV
Description	Aircraft engine with 8-inline cylinders, aluminium alloy crankcase with 9 main bearings, steel-made cylinder with thin-wall welded cooling jackets, also named Mercedes-type cylinder (Fig. 3.4)
Stroke	160 mm
Bore	140 mm
Displacement	19.77 L
Compression ratio	4.6
Power	232 hp at 1440 rpm (maximum output at sea level); 210 hp at 1300 rpm (max. service output at sea level)
Specific consumption	230 g/hp/h
Specific power	12.1 hp/litre
Valvetrain	2 overhead valves per cylinder operated by rocker arms by a single overhead camshaft
Fuel feeding	With two single-barrel petrol carburettors
Ignition	Two spark plugs per cylinder supplied by two high-voltage magneto Bosch HL8
Cooling	Liquid-fed by a centrifugal pump
Lubrication	Forced by a 4-cylinder reciprocating pump
Mass	410 kg
Mass-to-power ratio	1.71 kg/hp
Propeller transmission ratio	1.38 by means of a spur gears transmission
Application	AEG C.V (Fig. 3.5); AEG G.III; AGO C.II; Albatros C.V; Gotha G.II

Fig. 3.3 The aircraft engine Mercedes D.IV, 1915–1916

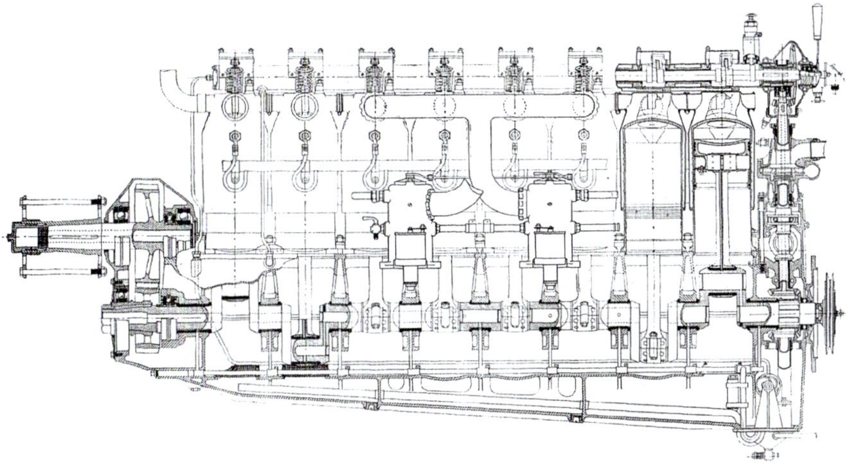

Fig. 3.4 Cross-section view of the aircraft engine Mercedes D.IV

Fig. 3.5 A reconnaissance biplane AEG C.V equipped with a Mercedes D.IV engine

3.3 Mercedes D.IVa

1916–1918

The Mercedes D.IVa was an aircraft engine with six-cylinder in line produced by the German company Daimler-Motoren-Gesellschaft in 1916 to replace the previous unreliable model Mercedes D.IV, in the range 200–299 hp ("klasse IV"). The D.IVa took up most of the technical features experienced by Mercedes, showing a new 4-valve per cylinder arrangement (Table 3.3) and a particular attention to frontal area reduction. To this purpose, the carburettor, normally mounted on right side, was placed behind the engine, and connected to the cylinders through a long intake manifold. This, in turn, entailed a not homogeneous supply of air–fuel mixture between the cylinders. The D.IVa was produced in different versions, for a total of 4550 units up to the end of the First World War. The engine was manufactured in two series, with different direction of rotation, for the use on multi-engine aircraft; later, to obtain greater power at high altitude, some supercharged versions were developed in cooperation with many companies, such as Schwade & Co, Brown Boveri and AEG. An experimental version with fuel injection was also developed in 1918 with Junkers.

Table 3.3 Technical specifications of the Mercedes D.IVa

Manufacturer	Daimler Motoren Gesellschaft, Stuttgart-Untertürkheim, Germany, 1916
Type	Mercedes D.IVa
Description	Aircraft engine with 6-inline cylinders, aluminium alloy crankcase with 7 main bearings, steel-made cylinder with thin-wall welded cooling jackets, also named Mercedes-type cylinder (Figs. 3.6 and 3.7)
Stroke	180 mm
Bore	160 mm
Displacement	21.7 L
Compression ratio	4.82
Power	260 hp at 1450 rpm (max. output at sea level)
Specific fuel consumption	230 g/hp/h
Valvetrain	2 overhead valves per cylinder operated by rocker arms by a single overhead camshaft
Fuel feeding	Single-barrel petrol carburettor
Ignition	Two spark plugs per cylinder supplied by two high voltage magneto Bosch Zh6
Cooling	Liquid-fed by a centrifugal pump
Lubrication	Forced by means of a 4-cylinder reciprocating pump
Mass	410 kg
Mass-to-power ratio	1.50 kg/hp
Application	Numerous military aircraft: AEG G.IV and R.I; Albatros C.X; Friedrichshafen G.III; Gotha G.III; Rumpler C.IV; Zeppelin-Staaken R.VI

Fig. 3.6 The crankshaft of the aircraft engine Mercedes D.IVa, 1916–1918

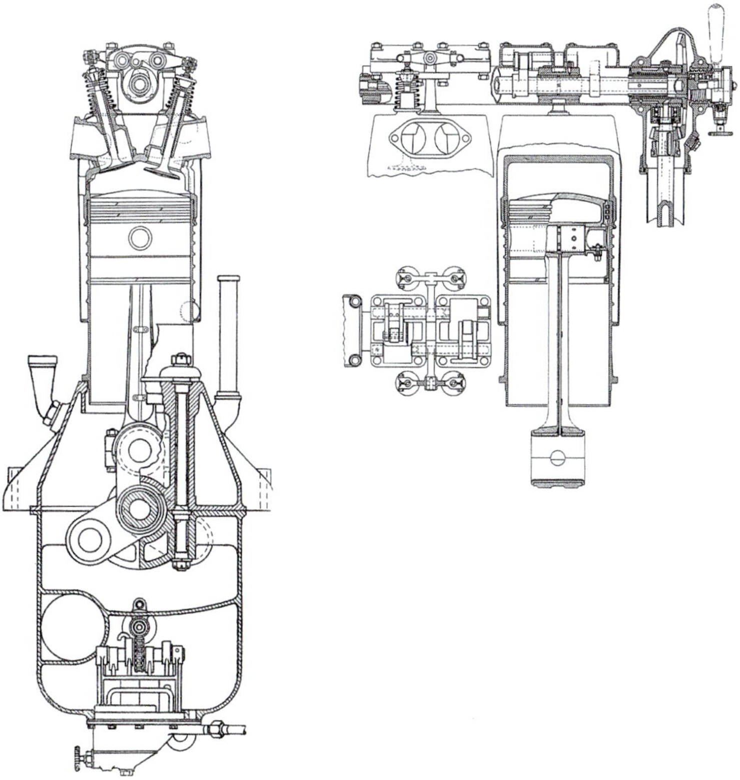

Fig. 3.7 Cross-section views of the aircraft engine Mercedes D.IVa, 1916–1918

3.4 Basse und Selve Bus.IV

1917–1918

Since 1906, Basse und Selve AG (B&S) was a famous factory with a leading position in the production of aluminium pistons and, similarly to many other companies, was involved in the military supply during the First World War for the German Idflieg (Inspektion der Fliegertruppen, Air Force Inspectorate). At the end of 1915, B&S overcame the scepticism of the military authorities on aluminium pistons reliability and was asked to develop a 6-inline-cylinder engine of 300 hp. The first 20-L version with 4 valves per cylinder, gave 260 hp and suffered several technical problems, forcing B&S to carry out a complete re-design. The new model, named Bus.IV (Figs. 3.8 and 3.9), featured a displacement of 22.7 L, and was successfully tested in October 1917 giving 278 hp in normal operating conditions (Table 3.4). B&S produced ca. 330 units of this engine to equip the heavier military aircraft, such as bombers and long-range reconnaissance aeroplane. The Bus.IV and the later models Bus.IVa and Bus.Vü were the greatest inline 6-cylinder engine built in series during the First World War.

Table 3.4 Technical specifications of the Basse und Selve Bus.IV

Manufacturer	Basse und Selve, Altena (Westfalia), Germany, from 1917 to 1918
Type	Bus.IV
Description	Aircraft engine with 6-inline cylinders, aluminium alloy crankcase with 7 main bearings, aluminium piston and steel-made cylinder with thin-wall welded cooling jackets, also named Mercedes-type cylinder
Stroke	200 mm
Bore	155 mm
Displacement	22.65 L
Compression ratio	4.7
Power	303 hp at 1450 rpm (maximum sea level power), 278 hp at 1400 rpm (sea level rating power)
Valvetrain	Four overhead valves per cylinder controlled by a single overhead camshaft with a desmodromic mechanism (Fig. 3.9)
Fuel feeding	Two single-barrel petrol carburettors
Ignition	Two spark plugs per cylinder supplied by two high voltage magneto Bosch Zh6
Cooling	Liquid-fed by a centrifugal pump
Lubrication	Forced by means of a 2-cylinder reciprocating pump
Mass	410 kg
Mass-to-power ratio	1.71 kg/hp
Application	Military aircraft Rumpler C.IV, Siemens-Schuckert R.VIII, Hansa-Brandenburg W.29

Fig. 3.8 The engine Basse und Selve Bus.IV, 1917–1918, probably one of very few survived

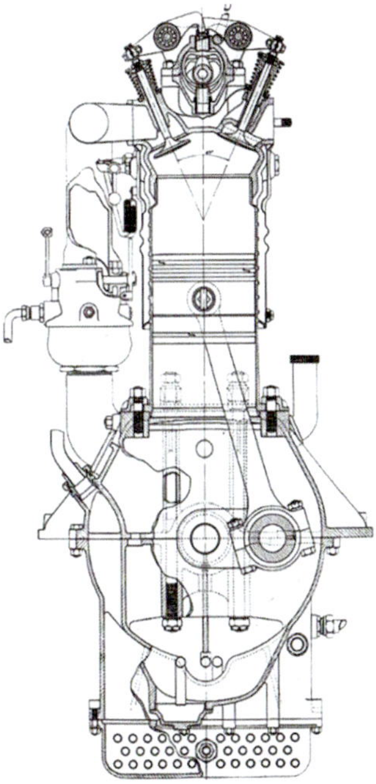

Fig. 3.9 Cross-section view of the aircraft engine Basse un Selve Bus.IV

3.5 Le Rhône Type J By

1917–1918

The aircraft engine Le Rhône Type J, produced by the French company Société des Moteurs Gnome et Rhône, has been widely used, since 1914, in numerous French and British aircrafts during the First World War. The same engine was manufactured under license in other countries: in Italy by the Società Italiana Motori Gnome e Rhône (SIMGER) in Turin to equip Nieuport aircrafts built under license by Macchi for the Italian Air Force Corps. Until the end of the First World War the engine underwent a continuous development, resulting in various upgraded models; the last and most powerful one was the Type J By (Table 3.5), equipped with aluminium pistons (Fig. 3.10). Basically, the Le Rhône Type J is typical a rotary engine, in which the crankshaft is fixed to the aircraft frame and the cylinder-block rotates around it, directly driving the propeller (Fig. 3.11). This arrangement, introduced in aviation in 1907 by the French brothers Luis and Laurent Seguin with their Gnôme Omega engine, was widely used from 1908 to 1918 by many manufacturers. It allowed to solve various matters of the early aircraft engines, above all a proper cooling effect. At the beginnings of the 1920s it resulted outdated by the development of other "conventional" engine designs.

Table 3.5 Technical specifications of the Le Rhone Type J By

Manufacturer	Société des Moteurs Gnome et Rhône, Paris, France, from 1917 to 1918
Type	J By
Description	9-cylinder radial rotary engine, crankcase and cylinders made of steel with aluminium pistons
Stroke	170 mm
Bore	112 mm
Displacement	15.07 L
Compression ratio	5.0
Power	130 hp at 1350 rpm (maximum output at sea level)
Specific fuel consumption	375 g/hp/h
Valvetrain	2 overhead valves per cylinder both operated by a single rocker arm and pushrod (for each cylinder)
Fuel feeding	One single-barrel petrol carburettor controlled by a sliding throttle
Ignition	One spark plug per cylinder supplied by two high voltage magneto
Cooling	Air-fed
Lubrication	Forced by means of a gear pump
Mass	132 kg
Mass-to-power ratio	1.0 kg/hp
Application	Numerous military aircraft: SPAD Herbemont; Hanriot HD.1; Morane-Saulnier AC; Nieuport 23, 24 e 27

Fig. 3.10 The aircraft engine Le Rhône Type J By, 1917–1918

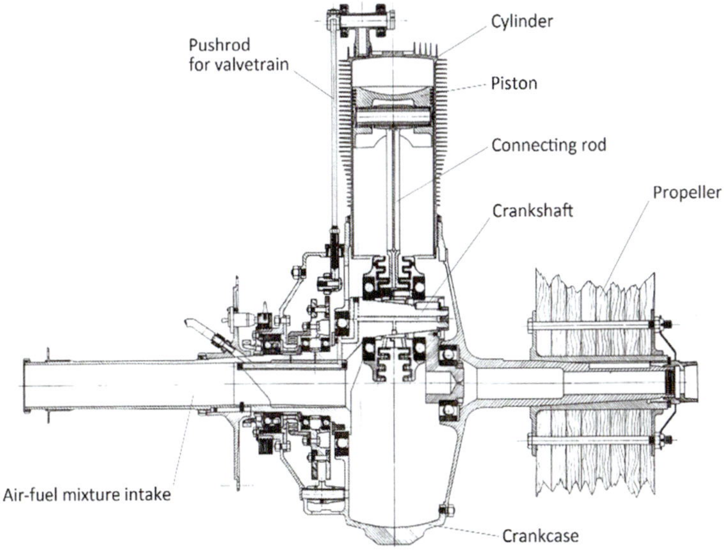

Fig. 3.11 Cross-section view of the aircraft engine Le Rhône Type J By

3.6 Siemens & Halske Sh.IIIa

1917–1918

The counter-rotary Siemens & Halske Sh.IIIa engine represents the latest development of the rotary engine arrangement (Figs. 3.12 and 3.13), widely used for aircraft propulsion in the early period of aviation, from 1908 to 1918. Despite of some drawbacks, the Sh.IIIa was a technically advanced engines for its time, capable of remarkable performances in terms of maximum power output, specific fuel consumption and mass-to-power ratio (Table 3.6). In September 1918, a Siemens-Schuckert D.IV equipped with the engine Sh.IIIa achieved a remarkable performance for those times, reaching the altitude of 8100 m from the ground in 36 min. Despite a total production of circa 2000 engines, manufactured by both Siemens-Halske and Rhemag Mannheim under license, only very few units remain nowadays.

Table 3.6 Technical specifications of the Siemens & Halske Sh.IIIa

Manufacturer	Siemens & Halske AG, Berlin-Spandau, Germany, from 1917 to 1918
Type	Sh.IIIa
Description	Counter-rotary radial 11-cylinder engine, in which cylinder rotate in opposite direction respect crankshaft by means of a bevel gear transmission, the propeller is directly driven by the cylinder block; crankcase and cylinders made of steel with aluminium pistons
Stroke	140 mm
Bore	124 mm
Displacement	18.6 L
Compression ratio	6.0
Power	160 hp at 900 rpm at sea level 200 hp at 950 rpm at sea level for take off/short time 200 hp at 1000 rpm at 2000 m (full throttle)
Specific fuel consumption	270 g/hp/h
Valvetrain	2 overhead valves per cylinder, operated by pushrods and rocker arms
Fuel feeding	One single-barrel petrol carburettor controlled by a sliding throttle
Ignition	One spark plug per cylinder supplied by two high-voltage magneto
Cooling	Air-fed
Lubrication	Forced by a 2-cylinder reciprocating pump
Mass	198 kg
Mass-to-power ratio	0.99 kg/hp
Application	Numerous military aircraft: Albatros D.XI, Pfalz D.VIII; Roland D.XVI, Siemens-Schuckert D.II, D.III and D.IV

Fig. 3.12 The aircraft engine Siemens & Halske Sh.IIIa, 1917–1918

Fig. 3.13 The aircraft engine Schematic cross-section view of the aircraft engine Siemens & Halske Sh.IIIa

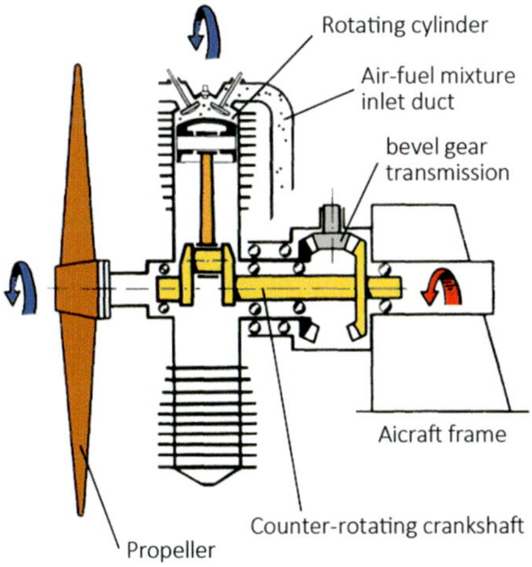

3.7 FIAT A 12 bis

1917–1919

In XX century, FIAT was one of the most important Italian car manufacturer, which started the autonomous production of both aircrafts and aero engines since the 1910s. The A 12 bis engine (Fig. 3.14) was developed in 1917 from the A 12 model, improving its overall performance and technical qualities. The engine has the typical technical features of liquid-cooled engines used in aviation during that period, whose design (Fig. 3.15) derived from a constructive arrangement introduced by Mercedes (Table 3.7). Cylinders and heads are made of light thin-walled weld-steel cooling jackets and are screwed to a common aluminium-alloy crankcase. This allowed to contain the total mass of the engine with a good reliability. Furthermore, various modifications could be easily introduced quickly, without substantial changes in the general layout. Produced in large series, for over 12,000 units, the FIAT A 12 bis was considered one of the best in its class and it was used, in the early 1920s, in numerous aircrafts in Italy as well as in French and Russia. In Great Britain it was sold at a higher price than some engines produced by other well-known companies and at the same price of the Rolls Royce Eagle, one of the most powerful of that period. After the war, some FIAT A 12 bis were also used in motorsport for speed record.

Table 3.7 Technical specifications of the FIAT A 12 bis

Manufacturer	FIAT, Torino, Italy, from 1917 to 1919
Type	A 12 bis
Description	Aircraft engine with 6-inline cylinders, aluminium alloy crankcase with 7 main bearings, aluminium pistons and steel-made cylinders with thin-wall welded cooling jackets (Mercedes-type cylinder)
Stroke	180 mm
Bore	160 mm
Displacement	21.7 L
Compression ratio	4.7
Power	300 hp at 1600 rpm at the sea level
Specific fuel consumption	225 g/hp/h
Valvetrain	Four overhead valves per cylinder controlled by a single overhead camshaft
Fuel feeding	Two single-barrel petrol carburettors
Ignition	Two spark plugs per cylinder supplied by two high voltage magnetos
Cooling	Liquid-fed by a centrifugal pump
Lubrication	Forced by gear pumps
Mass	438 kg
Mass-to-power ratio	1.5 kg/hp
Application	Numerous military aircraft: Ansaldo A.300; Caproni Ca.40; FIAT R.2; Macchi M.9

Fig. 3.14 The aircraft engine FIAT A 12 bis, 1917–1919

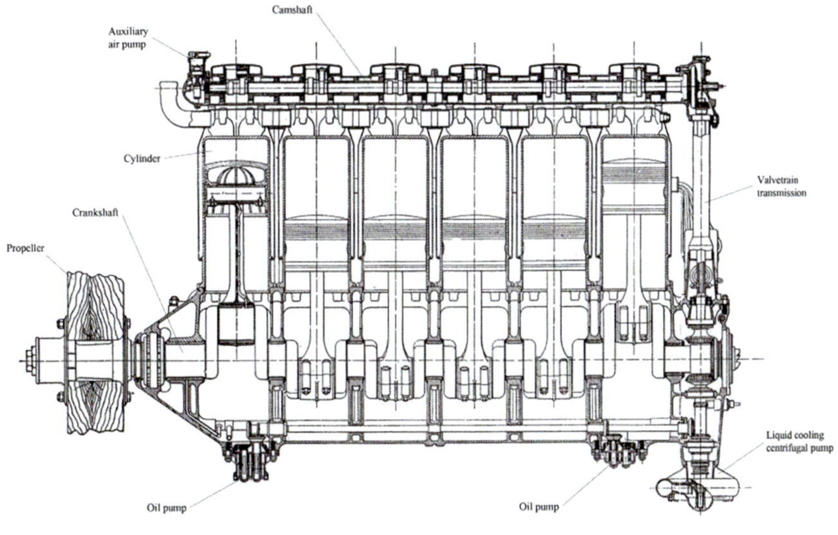

Fig. 3.15 Cross-section view of the aircraft engine FIAT A 12 bis

3.8 C.N.M.—Colombo D 110

1919–1923

At the beginning of the 1910s, the rapid development of the aeronautical sector prompted various companies to engage in the development of new engines specifically designed for aircraft and airships. Similarly to FIAT and SPA, the Officine Meccaniche Colombo in Milan also began the study of a liquid-cooled engine with six-in-line cylinders. Named D110 (Fig. 3.16), its production started in 1916 and it was also built under license by other companies, including Alfa Romeo and Officine Meccaniche De Vecchi (later C.N.M.) in Milan. Its construction scheme takes up some technical features quite widespread at the beginning of the 1900s (Table 3.8), such as the use of groups of twin cylinders arranged in line and screwed over a single aluminium alloy base, with heads made in one piece with the corresponding cylinders. Unlike other engines of that period, the cooling jackets are made of thin-walled brass sheets, screwed (rather than welded) externally onto each set of cylinders. The D 110 engine offered performance and reliability in line with those of its direct competitors, including the FIAT A 10, and was used in various aircraft built in Italy during the First World War and in the following years, such as that on display.

Table 3.8 Technical specifications of the C.N.M.—Colombo D 110

Manufacturer	C.N.M. under license of. Officine Meccaniche Colombo, Milan, from 1919 to 1923
Type	D 110
Description	Aircraft engine with 6-inline cylinders, aluminium alloy crankcase with 4 main bearings, 3 single-piece cast iron cylinder-head units, with screwed thin-walled brass cooling jackets
Stroke	160 mm
Bore	120 mm
Compression ratio	5.0
Power	110 hp at 1350 rpm above sea level
Specific power	10.1 hp/litre
Valvetrain	2 overhead valves in each cylinder, controlled via rocker arms by a single overhead camshaft
Fuel feeding	Single-barrel petrol carburettor
Ignition	Two spark plugs per cylinder supplied by two high voltage magnetos
Cooling	Liquid-fed by a centrifugal pump
Lubrication	Forced by gear pumps
Mass	225 kg
Mass-to-power ratio	1.95 kg/hp
Application	Military aircraft: Farman HF.20, Farman MF.11 (Fig. 3.17), SAML-Aviatik A.3

Fig. 3.16 The aircraft engine C.N.M.—Colombo D 110, 1919–1923. Entrusted by Aeronautica Militare

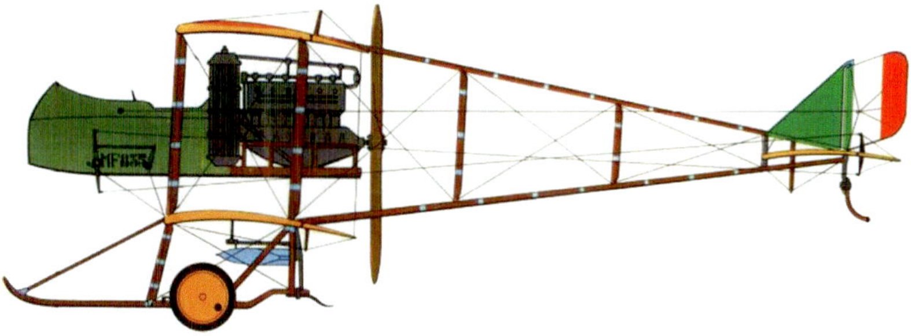

Fig. 3.17 The Farman MF.11 aircraft equipped with the Colombo D 110 engine

3.9 Colombo S. 53

1928–1930s

The S. 53 engine (Fig. 3.18) was designed and produced by the Officine Meccaniche Colombo in Milan, in 1928, for small trainer and general aviation aircraft. Subsequently the company was acquired by Alfa Romeo which continued the production of the engine until the end of the 1930s. Characterized by a simple construction scheme (Fig. 3.19), this engine has had a good commercial diffusion thanks to its ease of maintenance and its reliability. The four air-cooled steel cylinders are arranged in line, screwed over a common crankcase made of aluminium alloy (Table 3.9). Each of them has an independent aluminium alloy head. The propeller is directly driven by the crankshaft, without a speed reducer.

Table 3.9 Technical specifications of the Colombo S. 53

Manufacturer	Officine Meccaniche Colombo, Milan, Italy, from 1928 to 1930s
Type	S. 53
Description	Aircraft engine with 4-inline cylinders, aluminium crankcase with 5 main bearings, steel-made cylinders, head and pistons made of aluminium alloy
Stroke	140 mm
Bore	114 mm
Displacement	5710 cm^3
Compression ratio	5.24
Power	94 hp at 1900 rpm at sea level
Specific fuel consumption	230 g/hp/h
Specific power	16.5 hp/litre
Valvetrain	2 overhead valves per cylinder, operated by pushrods and rocker arms controlled by single camshaft housed in the crankcase
Fuel feeding	Single-barrel petrol carburettor Zenith 48 J 9, with an inlet mixture heater
Ignition	Two Marelli H4 spark plugs per cylinder supplied by two Marelli SA4 magnetos
Cooling	Air-fed
Lubrication	Forced by two gear pumps, one for pressure feeding and one scavenge
Weight	105 kg
Mass-to-power ratio	1.1 kg/hp
Application	Trainer and general aviation aircraft: Breda Ba.15.S, Breda Ba.39, Caproni Ca.125

Fig. 3.18 The aircraft engine Colombo S. 53, 1928–1930s

Fig. 3.19 Cross-section view of the aircraft engine Colombo S. 53

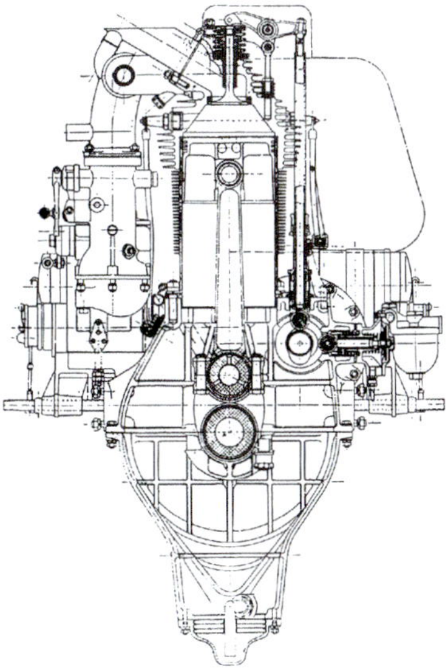

3.10 FIAT A 50

1928–1940

The FIAT A 50 (Fig. 3.20) is a 7-cylinder radial engine developed in 1928 for training and general aviation aircrafts. The engine features an ordinary single-row arrangement (Fig. 3.21) and thanks to a good reliability and an easy maintenance, it obtained a good market response. From 1928 to 1932, it was produced in circa 900 units which were fitted in many aircrafts up to 1940, such as the Caproni Ca.100 (Table 3.10). The A50 also achieved several aviation records in ceiling, distance and duration of flight. In 1930 the Italian aviator Francis Lombardi used an A 50 in his FIAT AS.1 e AS.2 aircrafts, for the long-distance flights Roma-Mogadiscio, Vercelli-Tokyo and for the circumnavigation of Africa. In the same period, a FIAT A 50 was used in one of the early helicopters of Corradino D'Ascanio, which achieved many flight records in its category.

Table 3.10 Technical specifications of the FIAT A 50

Manufacturer	FIAT, Turin, Italy, from 1928 to 1940
Type	A 50
Description	7-cylinder radial engine, crankcase, heads and pistons made of aluminium-alloy, streel-made cylinders screwed into the heads, crankshaft composed of two parts
Stroke	120 mm
Bore	100 mm
Displacement	6594 cm^3
Compression ratio	5.0
Power	100 hp at 1800 rpm at sea level power 55 hp at 1800 rpm at 5000 m above sea level
Specific fuel consumption	230 g/hp/h
Specific power	14.4 hp/litre
Valvetrain	2 overhead valves per cylinder, operated by pushrods and rocker arms using a wheel with a single row of cams, driven by an epicyclic gear train
Fuel feeding	Single-barrel petrol carburettor Zenith J 48
Ignition	Two Marelli S3 spark plugs per cylinder supplied by two Marelli MF7 magneto
Cooling	Air-fed
Lubrication	Forced by a gear pump
Weight	133 kg
Mass-to-power ratio	1.4 kg/hp
Application	Numerous trainer and general aviation aircraft: FIAT AS.1 and AS.2; Ambrosini SAI.3; CANSA C.5; CANT 26; Caproni Ca.100; IMAM Ro 5; D'AT3 (helicopter)

Fig. 3.20 The aircraft engine FIAT A 50, 1928–1940

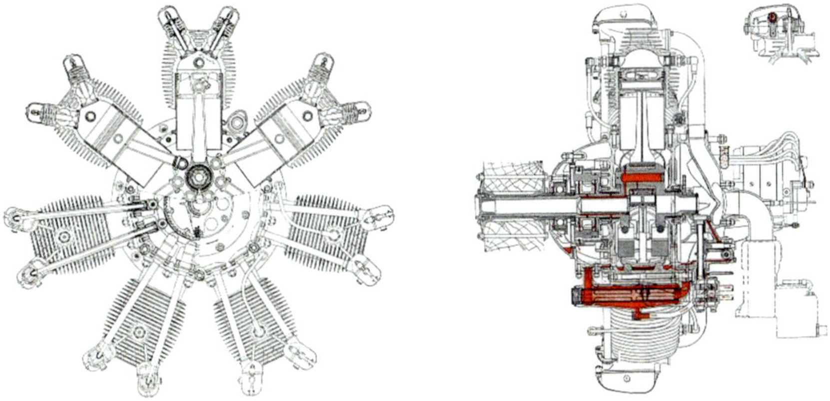

Fig. 3.21 Cross-section views of the aircraft engine FIAT A 50

3.11 Farina T 58

1930s

The T 58 (Fig. 3.22) is an aircraft engine for general aviation aircraft, designed and built in a few examples by Stabilimenti Farina, a well-known Turin company which, between 1906 and 1953, was particularly active in design and production of bodywork and mechanical systems for various automotive companies. The engine features a simple construction scheme (Table 3.11), with five radial cylinders, without a propeller speed reducer and supercharging system. Its structure and the arrangement of the main components, such as carburettor, magnetos and oil filter, are designed to ensure a very ease maintenance. Many parts, including the crankcase, are made of aluminium alloy. A refined system allows all the connecting rods to be connected directly to the single crankpin of the crankshaft instead of the more common arrangement with a primary connecting rod which drives the secondary ones. A limited production and a limited commercial spread contributed to making the T 58 engine very rare. That of the Museum is complete with every accessory and is also endowed with its original wooden transport case.

Table 3.11 Technical specifications of the Farina T 58

Manufacturer	S.A. Farina, Turin, Italy, 1930s
Type	T 58
Description	5-cylinder radial engine, crankcase, heads, and pistons made of aluminium-alloy, streel-made cylinders
Stroke	142 mm
Bore	120 mm
Displacement	8030 cm^3
Compression ratio	5.4
Power	140 hp at 1950 rpm, at sea level
Specific fuel consumption	210 g/hp/h
Brake mean effective pressure	8.1 bar
Specific power	17.4 hp/litre
Valvetrain	2 overhead valves per cylinder, operated by pushrods and rocker arms
Fuel feeding	Single-barrel petrol Zenith carburettor, equipped with an inlet air pre-heating system
Ignition	Two spark plugs per cylinder fed by two Marelli magnetos
Cooling	Air-fed
Lubrication	Forced by a gear pump
Mass	150 kg
Mass-to-power ratio	1.1 kg/hp
Application	Trainer and general aviation aircraft: Caproni Ca.100 (Fig. 3.23), Magni PM 3 / 4 Vale

Fig. 3.22 The aircraft engine Farina T 58, 1930s

Fig. 3.23 A seaplane Caproni Ca. 100 equipped with a Farina T 58 engine

3.12 Alfa Romeo 110 I

1933–1948

The 110 I (Fig. 3.24) is a small aircraft engine developed by Alfa Romeo in the early 1930s starting from the British de Havilland Gipsy Major engine. Approximately 500 units of it were produced until the end of the 1940s, in three construction series, in which various modifications and performance improvements were progressively introduced: the second version, known as the 110 bis, was produced from 1938 and the third one, the 110 ter1, in the postwar period. The 110 I has a simple construction layout (Fig. 3.25), quite widespread in the category of small and medium engine: it has four inverted in-line cylinders, without supercharging system and propeller speed reducer. Overall, the engine turned out to be very versatile, reliable, and easy to maintain. Thanks to these characteristics (Table 3.12) it was used successfully in many aircraft for long time.

Table 3.12 Technical specifications of the Alfa Romeo 110 I

Manufacturer	Alfa Romeo, Milan, from 1933 to 1948 (in three versions)
Type	110 I (first series)
Description	4 inline inverted-cylinder engine, aluminium alloy crankcase with 5 main bearings, independent steel cylinders with aluminium alloy heads
Stroke	140 mm
Bore	118 mm
Displacement	6.1 L
Compression ratio	5.25
Power	120 hp at 21,001 rpm at sea level
Specific power	19.6 hp/litre
Valvetrain	2 overhead valves in each cylinder, controlled via pushrods and rocker arms by a camshaft in the crankcase, driven by gears
Fuel feeding	One single-barrel petrol carburettor
Ignition	Two spark plugs per cylinder powered by two high voltage Marelli magnetos
Cooling	Air-fed
Lubrication	Forced by a gear pumps, one for pressure feeding and three for scavenge
Mass	136 kg
Mass-to-power ratio	1.13 kg/hp
Application	Various trainer and general aviation aircraft: SAI Ambrosini S.1001, Saiman 202, Breda Ba.79, FIAT G.2, Calzoni AC.240, CANT Z.1010 and Z.1012, Caproni 167

Fig. 3.24 The aircraft engine Alfa Romeo 110 I, 1933–1948. Entrusted by Aeronautica Militare

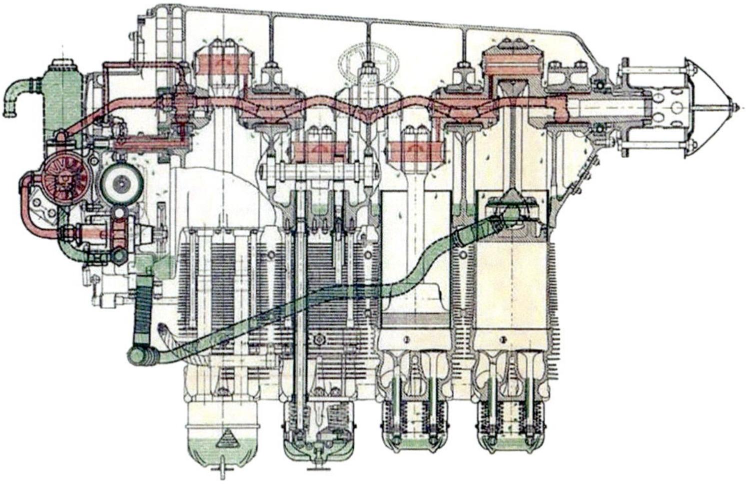

Fig. 3.25 Cross-section view of the aircraft engine Alfa Romeo 110 I

3.13 FIAT A 70

1934

In the first half of the 1930s FIAT intensified the study of air-cooled radial engines, developing various new models. Presented in 1934, the A 70 showed a whole of various refined technical characteristics, not common in small and medium-sized engines of that time. The A 70 (Fig. 3.26) features compact size and high-level constructive quality, observable in very thin cooling fins over both cylinders and heads, and good mass-to-power ratio, also obtained thanks the supercharging system (Table 3.13). Despite its excellent performance, improved in the more powerful A 70 S version in 1936, the engine gained no noticeable spread, and its production was limited to a few units.

Table 3.13 Technical specifications of the FIAT A 70

Manufacturer	FIAT, Torino, 1934
Type	A 70
Description	Single-row 7-cylinder radial engine, crankcase, heads and pistons made of aluminium-alloy, streel-made cylinders screwed into the heads, hemispherical combustion chambers
Stroke	115 mm
Bore	115 mm
Displacement	8.4 L
Compression ratio	5.75
Power	205 hp a 2200 rpm
Specific fuel consumption	250 g/hp/h
Specific power	24.5 hp/litre
Valvetrain system	2 overhead valves per cylinder, operated by pushrods and rocker arms
Fuel feeding	With one single-barrel gasoline FIAT carburettor, placed upstream of the compressor with air–fuel mixture preheating
Supercharging	One single-stage centrifugal compressor, driven by the crankshaft by means of a spur gears drive
Ignition	Two spark plugs per cylinder supplied by two independent high voltage magnetos
Cooling	Air-fed
Lubrication	Forced by gear pumps
Mass	167 kg
Mass-to-power ratio	0.81 kg/hp
Application	Various trainer and general aviation aircraft: FIAT G. 5 bis (Fig. 3.27), Breda 42, Cantieri Caproni Bergamasca PS. 1, Nardi FN 305 D

Fig. 3.26 The aircraft engine FIAT A 70, 1934. Entrusted by Aeronautica Militare

Fig. 3.27 The FIAT G. 5 aircraft, 1930s

3.14 FIAT A 74 R1C 38 *Ciclone*

1935–Early 1940s

In 1933 the General Staff of the Regia Aeronautica (Royal Italian Airforce) required the development of new high-performance air-cooled radial engines, preferring this arrangement respect to liquid-cooled one. To meet this requirements, the most important Italian engine manufacturers, such as FIAT and Alfa Romeo, purchased some manufacturing licenses from American and British companies in order to quickly build up under-license copies or redesigned model. To this purpose FIAT developed the A 74 (Figs. 3.28 and 3.29) engine for fighter and light bomber aircraft, as a result of redesign process based on both a Pratt and Whitney license and FIAT experience, carried out with previous own-design air-cooled engines (Table 3.14). The A 74 was manufactured up to the early 1940s, for a total production of circa 5500 units, including those made under license by Officine Meccaniche Reggiane. FIAT set up various versions, with specific settings and operating characteristics for different military aircraft during the Second World War.

Table 3.14 Technical specifications of the FIAT A 74 R1C 38 *Ciclone*

Manufacturer	FIAT, Turin, Italy, from 1935 up to early 1940s
Type	A 74 R1C 38, informally named *Ciclone* (Cyclone)
Description	Twin-row 14-cylinder radial engine, crankcase, heads and pistons made of aluminium-alloy, streel-made cylinders screwed into the heads
Stroke	145 mm
Bore	140 mm
Displacement	31.25 L
Compression ratio	6.7
Power	840 hp at 2400 rpm at 3800 m above sea level
Valvetrain	2 overhead valves per cylinder, operated by pushrods and rocker arms, sodium-cooled exhaust valves
Fuel feeding	One single-barrel petrol Zenith-Stromberg carburettor, placed upstream of the compressor
Supercharging	Single-stage centrifugal compressor, driven by the crankshaft by a spur gears drive (speed ratio 8.78)
Ignition	Two spark plugs per cylinder supplied by two independent high-voltage Marelli magnetos
Cooling	Air-fed
Lubrication	Forced by gear pumps
Weight	590 kg
Mass-to-power ratio	0.70 kg/hp
Application (regardless specific settings)	Military aircraft: FIAT C.R.42, FIAT G.50, FIAT C.R.25, FIAT G.12, FIAT CANSA F.C.20, FIAT CMASA R.S.14, Caproni Vizzola F.5, IMAM Ro.51, IMAM Ro.57, Macchi M.C.200 Saetta

Fig. 3.28 The aircraft engine FIAT A 74 R1C 38 *Ciclone*, 1935–early 1940s

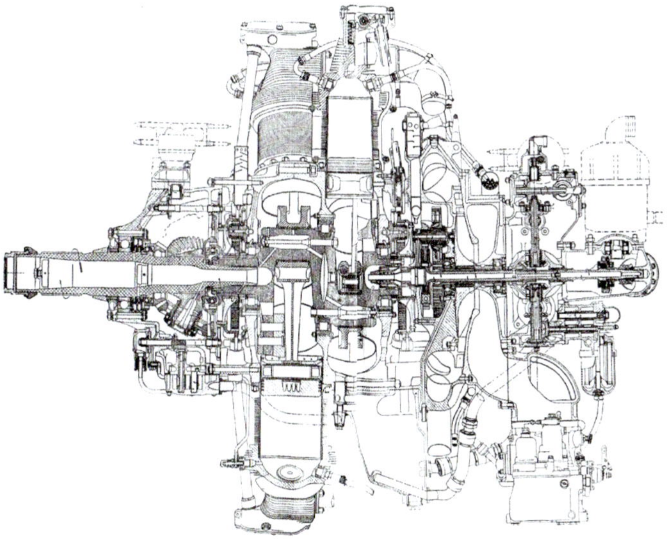

Fig. 3.29 Cross-section view of the aircraft engine FIAT A 74, showing the crankshaft composed of two parts connected by bolts, the hemispherical combustion chambers, and the Farman-type bevel gear epicyclic transmission for propeller speed reduction (ratio 0.725)

3.15 FIAT A 80 RC 41 *Nembo*

1935–Early 1940s

The FIAT A 80 (Fig. 3.30) is a twin-row, radial aircraft engine whose technical development is strictly connected to the contemporary smaller model A 74 (Fig. 3.28). Despite the main difference on cylinder number, both engines were designed with the same bore and stroke dimensions, as well as a large quantity of interchangeable parts to simplify production, logistical management of supplies and operative maintenance. With a power output of 1000 hp, the A 80 (Table 3.15) was the most powerful air-cooled engine of FIAT production up to the beginning of the Second World War. It was manufactured in various versions, with specific settings and operating characteristics for different aircrafts. Originally conceived for military applications, before the war it was also used in some liner aircrafts (Fig. 3.31).

Table 3.15 Technical specifications of the FIAT A 80 RC 41 *Nembo*

Manufacturer	FIAT, Turin, Italy, from 1935 up to early 1940s
Type	A 80 RC 41, informally also named *Nembo* (Nimbus)
Description	Twin-row 18-cylinder radial engine, crankcase, heads and pistons made of aluminium-alloy, streel-made cylinders screwed into the heads
Stroke	165 mm
Bore	140 mm
Displacement	45.72 L
Compression ratio	6.70
Power	1000 hp at 2100 rpm, at 4100 m above sea level
Valvetrain	2 overhead valves per cylinder, operated by pushrods and rocker arms, sodium-cooled exhaust valves
Fuel feeding	One twin-barrel petrol FIAT carburettor, placed upstream of the compressor
Supercharging	One single-stage centrifugal compressor, driven by the crankshaft by means of a spur gears drive with a speed ratio of 9.37:1, wide-open-throttle altitude 4100 m
Ignition	Two spark plugs per cylinder supplied by two independent high-voltage Marelli MF 18 magnetos
Cooling	Air-fed
Lubrication	Forced by gear pumps
Weight	725 kg
Mass-to-power ratio	0.73 kg/hp
Propeller speed ratio	0.625:1 by means of a Farman-type bevel gear epicyclic speed reducer
Application (regardless specific settings)	Military and liner aircraft: FIAT B.R.20, FIAT G.18 V, Breda Ba.65, Breda Ba.82, CANT Z.509 (Fig. 3.31), Caproni Ca.135, Savoia Marchetti S.M.79B (few units)

Fig. 3.30 The aircraft engine FIAT A 80 RC 41 *Nembo*, 1935–early 1940s

Fig. 3.31 A CANT Z.509 seaplane powered by three FIAT A 80 engines at the end of 1930s

3.16 Daimler-Benz DB 605

1941–1945

The DB 605 (Fig. 3.32) is a high-performance inverted V-12-cylinder engine developed by Daimler Benz in the early 1940s, for various types of military aircrafts. Strictly derived from the previous DB 601 model, it can be considered as one of the best aircraft piston-engine developed in Germany up to the end of the Second World War: this is due to several advanced technical features (Table 3.16), such as a Bosch in-cylinder direct injection and a single-stage compressor driven by hydraulic coupling, with a variable speed ratio controlled by an automatic control system. The DB 605 was manufactured in circa 42,400 units, considering all series from 1941

Table 3.16 Technical specifications of the Daimler-Benz DB 605

Manufacturer	Daimler-Benz AG, Stuttgart-Untertürkheim, Germany, from 1941 to 1945
Type	DB 605 A
Description	Inverted V-60° 12-cylinder engine, crankcase, heads and pistons made of aluminium-alloy, streel-made cylinders screwed into the heads
Stroke	160 mm
Bore	154 mm
Displacement	35.7 L
Compression ratio	7.3 on right bank, 7.5 on left bank
Power	1475 hp at 2800 rpm at sea level (for take-off)
Valvetrain	Four valves per cylinder, operated by single overhead camshaft in each bank, sodium-cooled exhaust valves
Fuel feeding	In-cylinder direct injection supplied by a Bosch petrol high pressure reciprocating pump
Supercharging	One single-stage centrifugal compressor driven by the engine crankshaft via a hydraulic coupling with continuous variable speed ratio (Fig. 3.33)
Ignition	Two spark plugs per cylinder supplied by two independent high-voltage Bosch magnetos
Cooling	Liquid-fed by a centrifugal pump, with a 50% mixture of water and ethylene glycol
Lubrication	Forced by feeding and scavenge gear pumps
Weight	720 kg
Mass-to-power ratio	0.49 kg/hp
Propeller speed ratio	1.6:1 by means of a spur gears transmission
Application (regardless specific settings)	Numerous military aircraft: Messerschmitt Me 109 G, Me 110 G/H and Me 210 C; Saab 18 and 21; FIAT G. 55, Macchi M.C. 205V, Reggiane RE. 2005 et al.

to 1945. It was also produced under license by other companies, such as in Italy by FIAT from 1942, for a total production of 1750 units, named RA 1050 RC 58I *Tifone*.

Fig. 3.32 The aircraft engine Daimler-Benz DB 605, 1941–1945

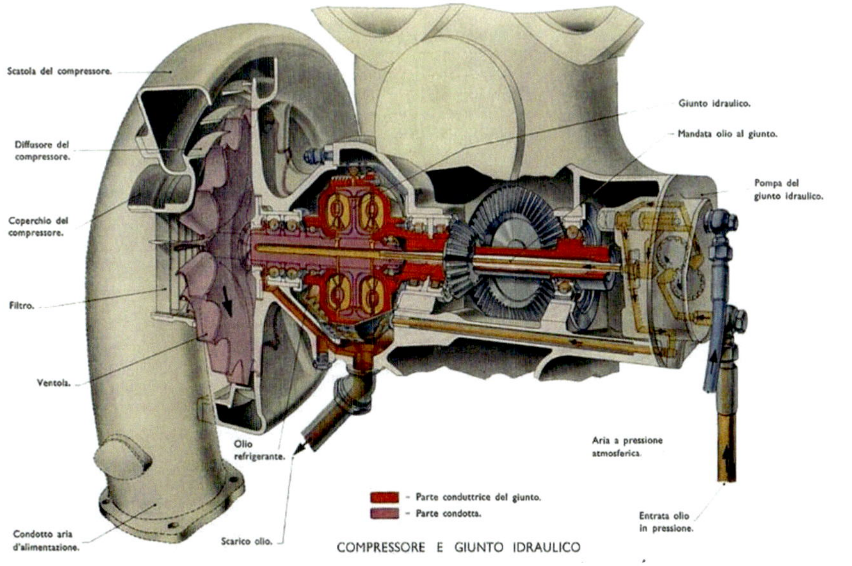

Fig. 3.33 Schematic cross-section view of the turbocharger and its hydraulic continuous variable transmission of an aircraft engine Daimler-Benz DB 605

3.17 Junkers Jumo 211 J-2

1942–1944

The Jumo 211 (Fig. 3.34) was one of the most successful and widespread Junkers aircraft engines. Its prototype was built in the mid-1930s starting from the 210 model to obtain better performance, as requested by the German Air Ministry, maintaining the same layout with inverted V-twelve cylinders and the sophisticated direct fuel injection system (Table 3.17). Launched in 1937, it was built in various factories, reaching over 68,000 units in 1944. Junkers built many versions of the 211 engine, introducing various technical and performance improvements: the 1942 model J, is one of the most powerful of those produced in series, which is endowed with an inter-cooler to lower the temperature of the compressed air supplied by the supercharger. Most of the Jumo 211 engines were used for military aircraft, including the Junkers

Table 3.17 Technical specifications of the Junkers Jumo 211 J-2

Manufacturer	Junkers Flugzeug und Motorenwerke, Germany from 1942 to 1944
Type	Jumo 211 J-2
Description	Inverted V-60° 12-cylinder engine, crankcase, heads and pistons made of aluminium-alloy (Fig. 3.35)
Stroke	165 mm
Bore	150 mm
Displacement	35 L
Compression ratio	6.5
Power	1420 hp at 2600 rpm at sea level for take-off
Valvetrain	Three overhead valves per cylinder, two for intake and one for exhaust, controlled via rocker arms by single overhead camshaft in each bank, sodium-cooled exhaust valves
Fuel feeding	In-cylinder direct injection supplied by a Junkers petrol high pressure reciprocating pump
Supercharging	One single-stage centrifugal compressor, driven by the crankshaft by means of a two-speed gearbox
Ignition	Two spark plugs per cylinder supplied by two independent high-voltage magnetos
Cooling	Liquid-fed by a centrifugal pump
Lubrication	Forced by feeding (1) and scavenge (3) gear pumps
Mass	720 kg (dry, without dynamo and inertial starter)
Mass-to-power ratio	0.51 kg/hp
Propeller speed ratio	0.530 by means of a spur gears transmission
Application	Military aircraft: Junkers Ju 87 D Stuka (Sturzkampfflugzeug) and Junkers Ju 88 A-4

Ju 87 Stuka (abbreviation for Sturzkampfflugzeug, i.e., dive bomber aircraft), also used in other countries, such as in Italy, during the Second World War.

Fig. 3.34 The aircraft engine Junkers Jumo 211 J-2, 1942–1944. Entrusted by Aeronautica Militare

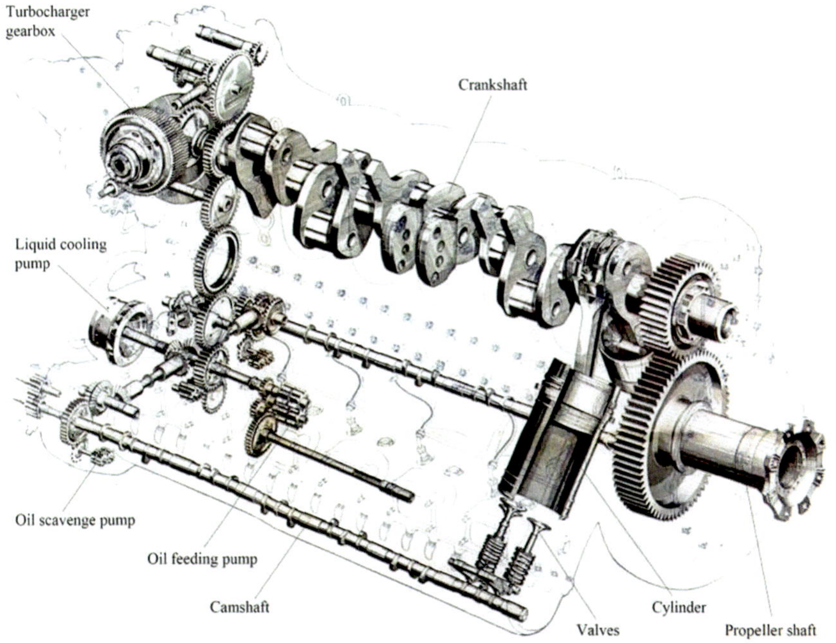

Fig. 3.35 Schematic cross-section view of the aircraft engine Junkers Jumo 211 J-2

3.18 Isotta Fraschini Beta RC.10 IZ

1940s

The Beta is an engine conceived in the early 1940s for general aviation aircraft, as well as advanced trainers with aerobatic capabilities. It belongs to a family of Isotta Fraschini engines introduced in the 1930s and characterized by a particular arrangement with inverted air-cooled cylinders. The Beta (Fig. 3.36) was developed from the larger V-12-cylinder Gamma and Delta models. Respect them it represented the smallest version, with only 6-in-line inverted cylinders (Table 3.18), proposed for an application range up to 300 hp, which was poorly manned by competitors at the time. The engine shares many components with the larger Gamma model, including cylinders, heads, pistons, and some parts of valvetrain (Fig. 3.37). The Beta was produced in the first half of the 1940s, but did not have much diffusion, due to some technical problems deriving from a quick development and to the consequences of the Second World War. At the end of 1948, Isotta Fraschini, already suffering for financial difficulties, terminated the production of aircraft engines. For these reasons, the overall production of all Beta engine was limited to a small number of units and those left to date are very rare.

Table 3.18 Technical specifications of the Isotta Fraschini Beta RC.10 IZ

Manufacturer	Isotta Fraschini Car Factory, Milan, 1940s
Type	Beta RC.10 IZ
Description	6-inline inverted-cylinder engine, aluminium alloy crankcase with independent cylinder-head groups
Stroke	130 mm
Bore	125 mm
Displacement	9.57 L
Compression ratio	6.5
Power	290 hp at 2550 rpm at 1000 m above sea level
Valvetrain	Two valves per cylinder and two overhead camshafts
Fuel feeding	Two single-barrel petrol carburettors placed downstream of the compressor
Supercharging	One single-stage centrifugal compressor, driven by the crankshaft by means of a single-speed transmission
Ignition	Two spark plugs per cylinder supplied by two independent high-voltage magnetos
Cooling	Air-fed
Lubrication	Forced by mechanical gear pump
Mass	277 kg
Mass-to-power ratio	0.96 kg/hp
Propeller speed ratio	0.774 by means of an epicyclic transmission
Application	General aviation and trainer aircraft: SAI Ambrosini S.7 (first units), Nardi FN316, CANSA C6B, IMAM Ro63, Breda 205, SAIMAN 208

Fig. 3.36 The aircraft engine Isotta Fraschini Beta RC.10 IZ, 1940s. Entrusted by Aeronautica Militare

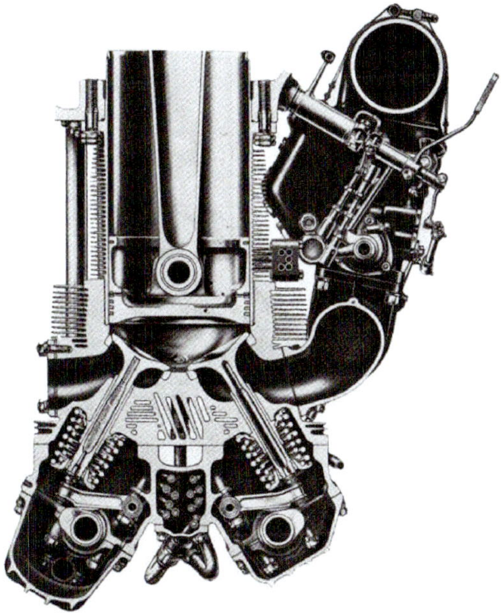

Fig. 3.37 Cross-section view of a cylinder-head assembly of the Isotta Fraschini Beta engine

3.19 Alfa Romeo 121 RC 20

1948

In the immediate post-war period various companies in the aeronautical sector attempted to resume industrial activities through the conversion of war production. The small and medium-power engines conceived for trainer aircraft between the second half of the 1930s and the early 1940s resulted in line with the new requirements of general aviation. To this end, Alfa Romeo resumed the 1943 project, creating a new model, named 121, with improved performance, thanks to a mechanically driven turbocharger and a propeller speed reducer (Table 3.19). The engine was produced in three versions, RC14, RC20 and RC22, with different altitude of maximum power output, respectively equal to 1400, 2000 and 2200 m above sea level. This was obtained through various settings of the supercharging system, according to the needs of specific aircraft. The inverted-cylinder arrangement with air-cooling system resulted in a compact and lightweight design, with performance comparable to that

Table 3.19 Technical specifications of the Alfa Romeo 121 RC 20

Manufacturer	Alfa Romeo, Milan, 1948
Type	121 RC 20
Description	Inverted V-90° 8-cylinder engine, aluminium alloy crankcase with independent cylinder-head groups
Stroke	110 mm
Bore	120 mm
Displacement	9.95 L
Compression ratio	6.5
Power	360 hp at 3000 rpm at 2000 m above sea level 410 hp at 3000 rpm above sea level (for take-off only)
Valvetrain	Two valves per cylinder, controlled by pushrods and rocker arms with a single camshaft for both banks
Fuel feeding	Single petrol carburettor upstream of the compressor
Supercharging	One single-stage centrifugal compressor, driven by the crankshaft by means of a single-speed transmission
Ignition	Two spark plugs per cylinder supplied by two independent high-voltage magnetos
Cooling	Air-fed
Lubrication	Forced by feeding (1) and scavenge (3) gear pumps
Mass	284 kg (dry)
Mass-to-power ratio	0.79 kg/hp
Propeller reduction ratio	0.621 by means of a spur gears transmission
Application	General aviation and advanced trainer aircraft SAI Ambrosini Super S.7 (Fig. 3.39, few units equipped)

of engines of the same category. Despite this, the production of the Alfa Romeo 121 (Fig. 3.38) was interrupted after a reduced number of units due to commercial failure and lack of production of the aircraft which it was designed for.

Fig. 3.38 The aircraft engine Alfa Romeo 121 RC 20, 1948. Entrusted by Aeronautica Militare

Fig. 3.39 One of the few SAI Ambrosini Super S.7 aircraft powered by Alfa Romeo 121 engine, displayed at the Museo Storico dell'Aeronautica Militare (Italian Air Force Museum) in Bracciano (Rome)

3.20 Blackburn Cirrus Bombardier

1950s

Designed in the late 1940s by the Cirrus Engine, a division of the British company Blackburn, the Bombardier (Fig. 3.40) is a small inverted-cylinder engine for light aircraft, produced in circa 300 units up to the mid-1950s, mainly to equip the reconnaissance aircraft Auster AOP.9. Despite its small size, the engine has fine technical characteristics (Table 3.20), such as a large use of light alloys, including the Elektron (a Magnesium alloy) for the crankcase and other structural parts. Moreover, it was one of the first British engines homologated for mass production to be equipped with a port-fuel-injection system, made by Skinner Union (SU). This allowed an increase in engine efficiency (i.e., a lower specific consumption), obtaining 25 hp more than the previous Cirrus Major III model with the same displacement.

Table 3.20 Technical specifications of the Blackburn Cirrus Bombardier

Manufacturer	Cirrus Engine, a division of the Blackburn and General Aircraft Ltd, Brough, Yorkshire (UK), 1950s
Type	Cirrus Bombardier
Description	4-inline inverted-cylinder engine, Elektron-made crankcase with 5 main bearings, steel-made cylinders, head and pistons in aluminium alloy (Fig. 3.41)
Stroke	140 mm
Bore	122 mm
Displacement	6.52 L
Compression ratio	7.0
Power	180 hp at 2600 rpm (max. output at sea level)
Specific fuel consumption	220 g/hp/h
Specific power	27.6 hp/litre
Valvetrain	2 valves per cylinder controlled by single overhead gear-driven camshaft
Fuel feeding	Port-injection supplied at 4.8 bar by a S.U. petrol reciprocating pump
Ignition	Two spark plugs per cylinder supplied by two high-voltage magnetos
Cooling	Air-fed
Lubrication	Forced by feeding and scavenge gear pumps
Length	1151 mm
Width	483 mm
Height	786 mm
Mass	159 kg
Mass-to-power ratio	0.88 kg/hp
Application	Various light aircraft: Auster AOP.9; Saunders-Roe Skeeter 3B; Miles M.38 Messenger

Fig. 3.40 The aircraft engine Blackburn Cirrus Bombardier, 1950s. Kind donation of Mr. Richard Webber with the support of Mr. Richard Horner from UK

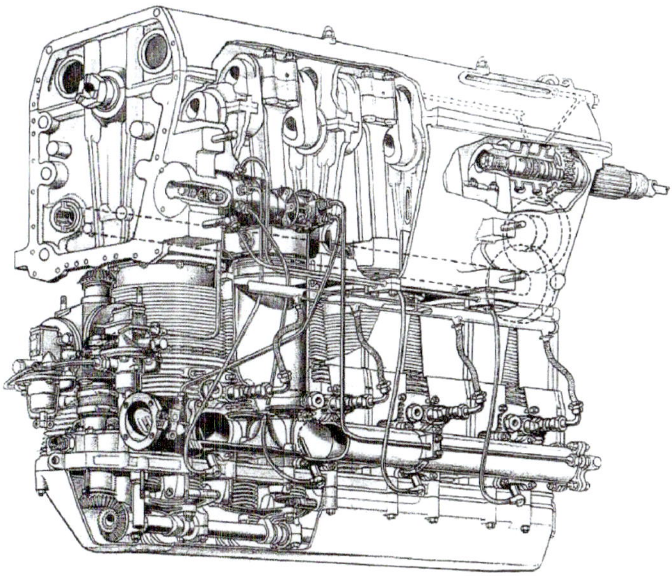

Fig. 3.41 Cross-section view of the aircraft engine Blackburn Cirrus Bombardier

3.21 Franklin 6A8

1945–1951

The Franklin 6A8 (Fig. 3.42) is an engine for light aircraft, such as trainer, built by Aircooled Motors Incorporated, starting from the second half of the 1940s. Its overall arrangement (Fig. 3.43), with 6 horizontally opposed (boxer) air-cooled cylinders, has been widely used up nowadays in aviation by many manufacturers, including Lycoming and Rotax. The 6A8 engine has been widely used for numerous aircraft around the world, such as the Republic RC-3 Seabee and the Piaggio P.136 seaplanes (Table 3.21). Thanks to its reliability and ease of maintenance, various units are still regularly used today, especially in the United States. The engine on display is complete with all accessories, including the oil, and has an original Piaggio pushing propeller, equipped with an in-flight variable pitch servomechanism.

Table 3.21 Technical specifications of the Franklin 6A8

Manufacturer	Aircooled Motors, Syracuse, New York, USA, from 1945 to 1951
Type	6A8-215-B9F
Description	6-opposed-cylinder engine (boxer-type), aluminium alloy crankcase composed of two parts with 5 main bearings, cylinders with integrated aluminium alloy head and internal hot-forced steel liners
Stroke	108 mm
Bore	127 mm
Displacement	8.21 L
Compression ratio	7.0
Power	215 hp at 2500 rpm
Valvetrain	2 overhead valves per cylinder, operated by pushrods and rocker arms
Fuel feeding	One single-barrel Marvel Schebler petrol carburettor
Ignition	Two spark plugs per cylinder supplied by two independent systems: an Auto-Lite inductive-discharge-type distributor and a Scintilla high-voltage magneto
Cooling	Air-fed, with a large fan directly driven by the crankshaft
Lubrication	Forced by a gear pump with oil cooler
Mass	196 kg (with cooling fan and oil cooler)
Mass-to-power ratio	0.91 kg/hp
Application	Various light aircraft: Piaggio P.136; Republic RC-3 Seabee; Waco Model W Aristocraft

Fig. 3.42 The aircraft engine Franklin 6A8 with a Piaggio two-blade variable-pitch propeller

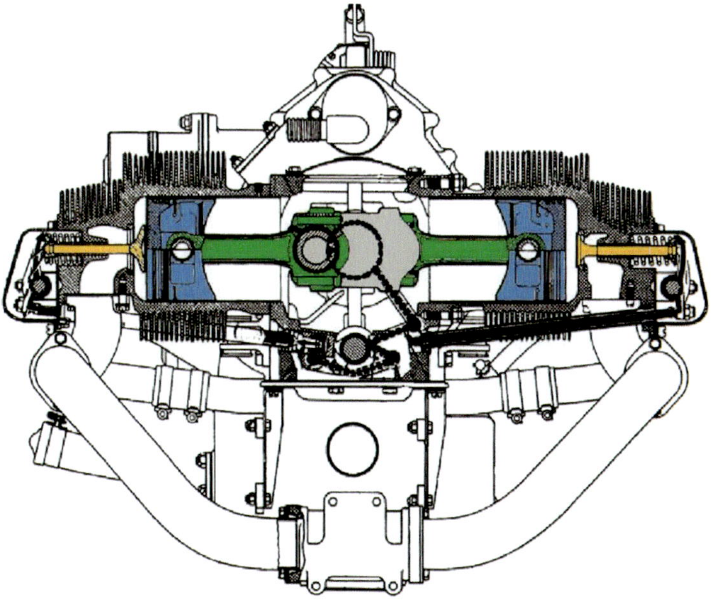

Fig. 3.43 Cross-section view of the aircraft engine Franklin 6A8

3.22 Franklin 6V4

1946–1970

Built by the American Aircooled Motors Incorporated, the Franklin 6V4 (Fig. 3.44) is an air-cooled 6-cylinder boxer engine specifically conceived for small helicopters. It was strictly derived from the 6A8 aircraft model (Fig. 3.42), with which it shares numerous components. Compared to the latter, the 6V4 engine is designed for vertical installation to facilitate the direct coupling with the helicopter's main rotor transmission (Table 3.22). Bell Helicopter Textron used this engine for some of its 47 series helicopters, which were among the first built for general aviation. In the 1950s, the Italian company Agusta acquired the license to produce the 47-G helicopters (Fig. 3.45) which were originally powered by the Franklin 6V4 engine.

Table 3.22 Technical specifications of the Franklin 6V4

Manufacturer	Aircooled Motors, Syracuse, New York, USA, from 1946 to 1970
Type	6V4-200-C32
Description	6-opposed-cylinder engine (boxer-type), aluminium alloy crankcase composed of two parts with 5 main bearings, cylinders with integrated aluminium alloy head and internal hot-forced steel liners
Stroke	89 mm
Bore	114 mm
Displacement	5.5 L
Compression ratio	8.5
Power	200 hp at 3100 rpm
Valvetrain	2 overhead valves per cylinder, operated by pushrods and rocker arms
Fuel feeding	One single-barrel Marvel Schebler petrol carburettor
Ignition	Two spark plugs per cylinder supplied by two independent Scintilla high-voltage magnetos
Cooling	Air-fed
Lubrication	Forced by a gear pump
Mass	135 kg (dry)
Mass-to-power ratio	0.76 kg/hp
Propeller speed ratio	9:1 by means of an epicyclic drivetrain
Application	Light helicopter: Bell 47 (series G, E, HTL-3 and H1)

Fig. 3.44 The aircraft engine Franklin 6V4, 1946–1970

Fig. 3.45 An Agusta-Bell 47-G equipped with a Franklin 6V4 engine

3.23 Curtiss-Wright R-1820 *Cyclone*

1930s–1963

The Curtiss-Wright Corporation is one of the most important US aviation companies. Founded in 1929 from the merger of Curtiss Airplane and Motor Company, Wright Aeronautical and other aviation companies, nowadays it operates in many other fields, such as aerospace, with a large production including aircraft and engines. From early 1930s its Cyclone-series engines became very spread and technically appreciated all over the world. Produced for over 30 years with a total of over 200,000 units, they were used all over the world in numerous military and civil aircraft up to the 1980s. This longevity was due to the layout of the original project (Table 3.23), which allowed a wide technical evolution: from the single-row 9-cylinder R-1820 up to the larger double-row models. Each of them was progressively developed in various sub-series with better performance, greater reliability, and service economy.

Table 3.23 Technical specifications of the Curtiss-Wright R-1820 *Cyclone*

Manufacturer	Curtiss-Wright Corporation, New Jersey, USA, from 1930s to 1963
Type	R-1820—1000 C9HD1, a late 1950s model (Fig. 3.46)
Description	Single-row 9-cylinder radial engine, aluminium alloy crankcase, steel-made cylinders, heads and pistons made of aluminium alloy (Fig. 3.47)
Stroke	174 mm
Bore	155.6 mm
Displacement	29.8 L
Compression ratio	6.8
Power	1425 hp at 2700 rpm at sea level
Valvetrain	2 overhead valves per cylinder controlled by pushrods and rocker arms, sodium-cooled exhaust valves
Fuel feeding	Single twin-barrel Stromberg petrol-injection carburettor placed upstream of the compressor
Supercharging	One single-stage centrifugal compressor, driven by the crankshaft by means of a two-speed gearbox
Ignition	Two spark plugs per cylinder supplied by two independent high-voltage magnetos
Cooling	Air-fed
Lubrication	Forced by feeding and scavenge gear pumps
Mass	630 kg
Mass-to-power ratio	0.44 kg/hp
Propeller reduction ratio	0.66 by means of an epicyclic drivetrain
Application (regardless specific settings)	Numerous military and civil aircraft: Boeing B-17; Douglas A-24, C-49, C-50; Grumman J2F-5; Lockheed A-29, C-60; North American O-47; Sikorsky H-34; Consolidated P2Y et al.

Fig. 3.46 The aircraft engine Curtiss-Wright R-1820 *Cyclone*, 1930–1963. Entrusted by Aeronautica Militare

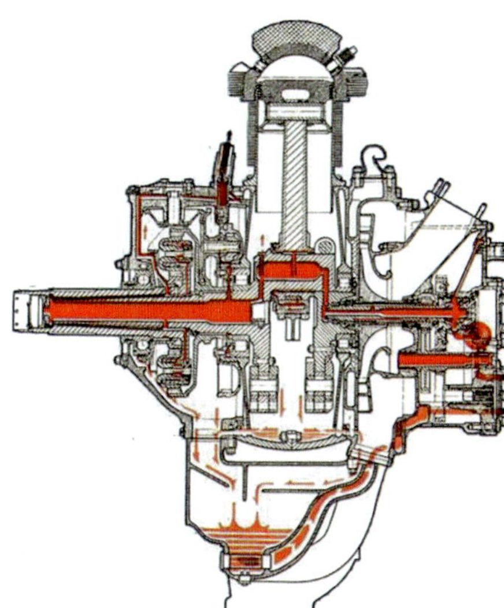

Fig. 3.47 Cross-section view of the aircraft engine Curtiss-Wright R-1820 *Cyclone*

3.24 Pratt and Whitney R-1830 *Twin Wasp*

1932–1951

Founded in 1925 and still active nowadays, the Pratt and Whitney Aircraft Company is one of the most important US companies in the aeronautical and aerospace sector. Its first engine, named R-1340 Wasp, was completed at the end of 1925, and became the archetypal model of a long series of engines that gained considerable commercial success all over the world thanks to their performance and high reliability. In particular, the R-1830 model, developed in the early 1930s, is the most famous of the Wasp series as well as one of the most produced piston engines in the history of aviation, together with its counterpart and direct competitor, the Wright R-1820. The R-1830 engine was produced up to 1951 in over 170,000 units for many military and civil aircraft. Its design (Figs. 3.48 and 3.49), almost unchanged over the years, allowed the development of the original project through dozens of models

Table 3.24 Technical specifications of the Pratt and Whitney R-1830 *Twin Wasp*

Manufacturer	Pratt and Whitney Aircraft, USA, from 1932 to 1951
Type	R-1830/92—S1C3G, a late 1940s model (Fig. 3.48)
Description	Twin-row 14-cylinder radial engine, aluminium alloy crankcase, steel-made cylinders with forged pistons and heads (screwed) made of aluminium alloy
Stroke	139.5 mm
Bore	139.5 mm
Displacement	29.85 L
Compression ratio	6.7
Power	1200 hp at 2700 rpm at sea level
Valvetrain	2 overhead valves per cylinder controlled by pushrods and rocker arms, sodium-cooled exhaust valves
Fuel feeding	Twin-barrel Stromberg petrol-injection carburettor placed upstream of the compressor
Supercharging	One single-stage centrifugal compressor, driven by the crankshaft through a single-speed gearbox (ratio 7.15)
Ignition	Two spark plugs per cylinder powered by two independent high-voltage Scintilla magnetos
Cooling	Air-fed
Lubrication	Forced by feeding and scavenge gear pumps
Mass	662 kg
Mass-to-power ratio	0.55 kg/hp
Propeller reduction ratio	0.5625 by means of a bevel gear epicyclic drivetrain
Application	Various aircraft: Douglas C-47, C-52, Lockheed A-28; Consolidated B-24; Curtiss P-36; Republic P-43 et al.

and various sub-types. They were subject to continuous technical updates and great performance improvements, ranging from circa 650 hp of the first prototypes up to over 1350 hp of the latest models, with almost unchanged original layout and dimensions (Table 3.24).

Fig. 3.48 The aircraft engine Pratt and Whitney R-1830 *Twin Wasp*, 1932–1951. Entrusted by Aeronautica Militare

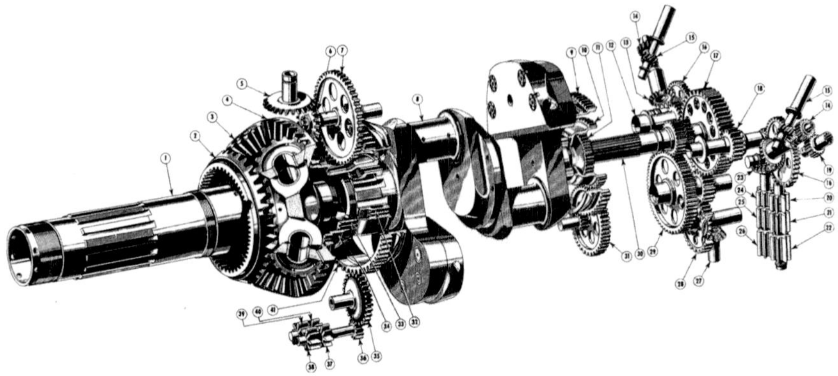

Fig. 3.49 The powertrain components of the Pratt and Whitney R-1830; from left to right: the propeller shaft with the Farman-type bevel-gear epicyclic speed reducer, the crankshaft with its balancing counterweights, and the accessories drivetrain

3.25 Alvis Leonides 502

1947–1958

In the mid-1930s, the English company Alvis, already operating in the automotive and aeronautical fields, began the development of a new 9-cylinder radial engine, completing its first prototype in 1936. After various flight tests carried out before the Second World War, due to the logistical priorities imposed by the conflict, the production of the engine began only in 1947. Under the name Leonides, two different series were developed: the 500-series engines for aircraft (Table 3.25) and the 520-series units for helicopters, adding particular accessories such as a large crankshaft-driven cooling fan. Alvis developed various models for each series, with specific characteristics and performance for numerous aircraft. The Leonides remained in production up to 1966 and was the last high-power piston aero engine to be produced in the United Kingdom. In Italy some engines of the 500 series (Fig. 3.50) were used in the prototype of the four-engine Agusta A.Z.8 in 1958 (Fig. 3.51) and in the first prototype of the trainer aircraft FIAT G. 49 in 1953.

Table 3.25 Technical specifications of the Alvis Leonides 502

Manufacturer	Alvis, Coventry, UK, from 1947 to 1958
Type	502/5
Description	Single-row 9-cylinder radial engine, aluminium alloy crankcase, steel-made cylinders with forged pistons and heads (screwed) made of aluminium alloy
Stroke	112 mm
Bore	122 mm
Displacement	11.8 L
Compression ratio	6.8
Power	520 hp at 3000 rpm
Valvetrain	2 overhead valves per cylinder controlled by pushrods and rocker arms, sodium-cooled exhaust valves
Fuel feeding	Hobson petrol-injection carburettor placed upstream of the compressor
Supercharging	One single-stage centrifugal compressor, driven by the crankshaft through a single-speed gearbox (ratio 6.5)
Ignition	Two spark plugs per cylinder powered by two independent high-voltage B.T.H. magnetos
Cooling	Air-fed
Lubrication	Forced by feeding and scavenge gear pumps
Mass	363 kg
Mass-to-power ratio	0.70 kg/hp
Propeller reduction ratio	0.625 by means of a bevel gear epicyclic drivetrain
Application	Various aircraft: Percival P.50 and P.66; Scottish Aviation Pioneer; de Havilland Canada DHC-2 Beaver Mk.2; Agusta A.Z.8; FIAT G. 49–1 (prototype)

Fig. 3.50 The aircraft engine Alvis Leonides 502, 1947–1958. Entrusted by Aeronautica Militare

Fig. 3.51 Advertisement of Alvis Leonides engines with their application for the Agusta A.Z.8

3.26 De Havilland Goblin

1940s–Early 1950s

The Goblin (Fig. 3.53) was one of the first turbojet engines to be used in experimental aircraft during World War II and then in the first mass-produced jet aircraft. The original prototype, called H-1, was developed in the early 1940s by Frank Halford, who later designed a larger model named Ghost. Both the engines have the same arrangement (Fig. 3.54), mainly used by British manufactures up to 1950s: a single-stage centrifugal compressor driven by a single stage axial-flow turbine, with direct-flow can-type combustion chambers. Initially the Goblin was mass-produced by the de Havilland Engine Company, later by Rolls-Royce when it acquired the first one (Table 3.26).

Table 3.26 Technical specifications of the de Havilland Goblin

Manufacturer	De Havilland Engine Co., Watford, United Kingdom, from 1940s to early 1950s
Type	Goblin 3
Description	Single-shaft turbojet engine (Fig. 3.54)
Compressor	Single-stage centrifugal-flow, forged aluminium alloy impeller with 17 blades
Compression ratio	3.67
Air mass flow	28 kg/s at 10,750 rpm at sea level in static conditions
Combustor	16 direct-flow can-type chambers, made of sheet steel
Turbine	Single-stage axial-flow, rotor with 93 blades
Exhaust nozzle	Fixed geometry cone
Fuel feeding	Two variable-stroke reciprocating jet fuel pumps, with barometric control of air/fuel mixture
Lubrication	Forced by feeding and scavenge gear pumps
Max. static thrust	14.7 kN (see air mass flow)
Diameter	1266 mm
Mass	735 kg
Mass/Thrust ratio	0.50 kg/daN
Time between overhaul	250 h
Application (regardless specific settings/model)	Various aircraft: de Havilland DH.100 Vampire (Fig. 3.52), Saab 21R, FIAT G. 80

Fig. 3.52 Transparent view of a de Havilland DH.100 Vampire with Goblin turbojet engine

Fig. 3.53 The aircraft engine de Havilland Goblin, 1940s–early 1950s. Entrusted by Aeronautica Militare

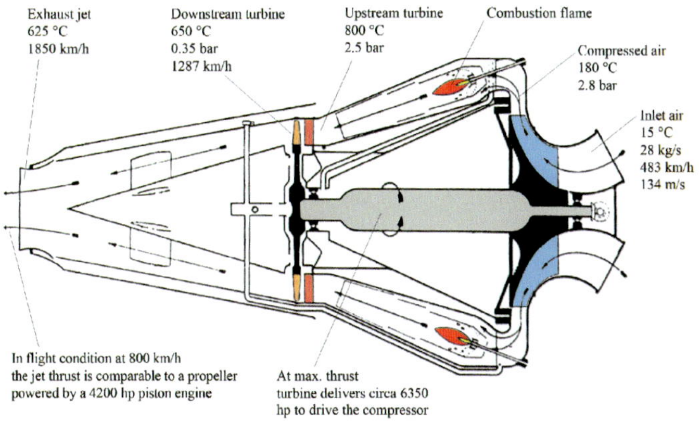

Fig. 3.54 Schematic cross-view of the aircraft engine De Havilland Goblin

3.27 De Havilland Ghost

1940s–Late 1950s

The de Havilland Ghost was one of the first turbojet engine to be developed in United Kingdom starting from the last years of World War II. It derives from the Frank Halford's H-2 prototype, conceived from the smaller turbojet H-1 to achieve the performance requirements for the de Havilland Comet, the first commercial airliner with turbojet engines. During the development process, the de Havilland acquired the Halford's company and renamed the prototypes H-1 and H-2 as Goblin and Ghost respectively. The first tests on the Ghost engine started in 1944 and as a result it was ready in 1945, with a certain advance respect the Comet aircraft. Meanwhile, it was used to power the Swedish fighter Saab J29 Tunnan, who made his first flight in 1948. In 1949, the Ghost was finally used for the Comet, powered with four engines, and then for the fighter-bomber de Havilland Venom. In the following years various versions were developed for higher performance until the end of production in late 1950s. The Ghost was buit under license by Svenska Flygmotor (later Volvo Aero) for the Saab J29 Tunnan and by FIAT (48-1 version) for a brief period (Table 3.27).

Table 3.27 Technical specifications of the De Havilland Ghost

Manufacturer	De Havilland Engine Co., Watford, United Kingdom, from 1945 to late 1950s; FIAT Aviation under license, in the early 1950s
Type	Ghost
Description	Single-shaft turbojet engine
Compressor	Single-stage centrifugal-flow, aluminium alloy forged impeller with 19 blades (Fig. 3.55)
Compression ratio	4.5
Air mass flow	40 kg/s at 10,250 rpm at sea level in static conditions
Combustor	8 direct-flow can-type chambers, steel-sheet made
Turbine	Single-stage axial-flow, rotor with 97 blades made of Nimonic 90 alloy, both sides of the rotor cooled by air bled from compressor; inlet gas temperature 800 °C, outlet temperature 650–700 °C at 10,250 rpm
Exhaust nozzle	Fixed geometry cone (Fig. 3.56)
Fuel feeding	Two variable-stroke reciprocating jet fuel pumps (in parallel) with barometric control of air/fuel mixture and 10 Lucas Duplex CSH-39 fuel injectors
Lubrication	Forced by feeding and scavenge gear pumps
Max. static thrust	22.0 kN at 10,250 rpm at sea level
Specific fuel consumption	1.07 kg/daN/h
Mass	996 kg
Mass/Thrust ratio	0.45 kg/daN
Application (regardless specific settings/ model)	Military and civil aircraft: de Havilland DH.106 Comet and DH.112 Venom; Saab J29 Tunnan

Fig. 3.55 The centrifugal compressor impeller (on the left) and the axial-flow turbine rotor (on the right) of a de Havilland Ghost turbojet engine, 1940s–late 1950s

Fig. 3.56 The aircraft turbojet engine de Havilland Ghost, 1940s–late 1950s

3.28 Rolls-Royce Derwent

1940s–Early 1950s

The Rolls-Royce Derwent (Fig. 3.57) was one of the first turbojet engines built in large series after World War II to be used for various military and civil aircraft. Derived from the 1943 Rolls-Royce Welland engine, the Derwent features the same overall layout designed by Frank Whittle, with a large single-stage centrifugal compressor (Fig. 3.58). Its technical development, started in 1944, was closely related to the larger and more powerful Rolls-Royce Nene. The Derwent remained in production up to the early 1950s in various upgraded models, with a continuous technical and performance improvement (Table 3.28). In Italy it was used to power the Aerfer Sagittario 2, the first Italian aircraft which exceeded the speed of sound on 4th December 1956.

Table 3.28 Technical specifications of the Rolls-Royce Derwent

Manufacturer	Rolls-Royce, Derby, UK, mid-1940s to early 1950s
Type	Derwent (data relating to version 5)
Description	Single-shaft turbojet engine (Fig. 3.58)
Compressor	Single-stage centrifugal-flow, double-entry impeller with 30 blades in each side, requiring circa 8000 hp in max. static thrust conditions
Compression ratio	4.1
Air mass flow	28.1 kg/s at 15,000 rpm at sea level
Combustor	10 direct-flow can-type chambers, steel-sheet made with swirl inducers around the fuel nozzles, max. admitted temperature 837 °C
Turbine	Single-stage axial-flow, forged-steel impeller with 54 Nimonic blades, air-cooled
Exhaust nozzle	Fixed geometry cone
Fuel feeding	One double flow injector (duple burner) in each combustion chamber, operated by an automatic control system on the basis of flight altitude and engine operating conditions (rotation speed, air pressure and temperature et al.)
Lubrication	Forced by gear pumps
Max. static thrust	17.8 kN at 15,000 rpm at sea level
Specific fuel consumption	1.04 kg/daN/h
Mass	581 kg (including electric starter and lubricating oil tank, excluding exhaust nozzle)
Mass/Thrust ratio	0.33 kg/daN
Application (regardless specific settings/ model)	Various military aircraft: Gloster Meteor F4 et al.; Avro Canada C-102 Jetliner; Fokker S.14; Aerfer Ariete and Aerfer Sagittario 2 (prototypes)

Fig. 3.57 The aircraft engine Rolls-Royce Derwent, 1940s–early 1950s. Entrusted by Aeronautica Militare

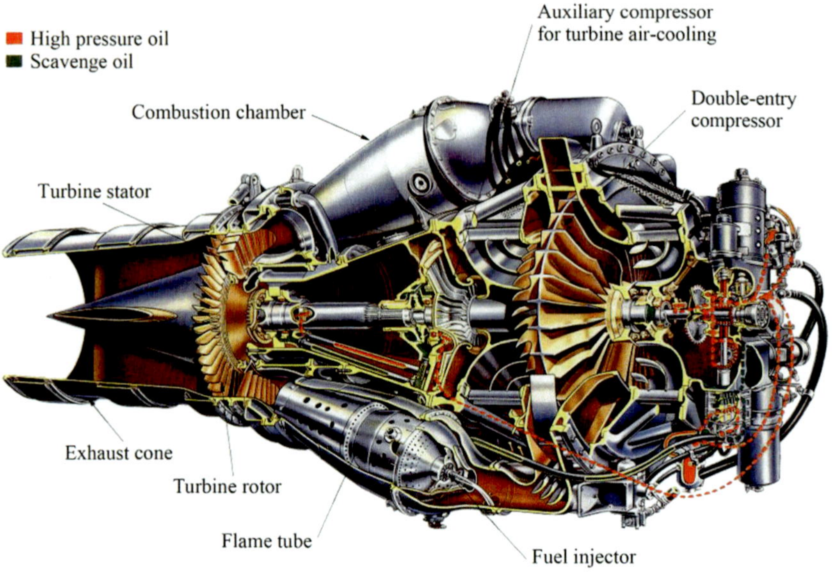

Fig. 3.58 Cross-section view of the aircraft engine Rolls-Royce Derwent

3.29 Allison J33 Turbine

1945–1950s

The Allison J33 was one of the first jet engines to be mass-produced in the United States in the 1940s. Its development process was initially conducted by General Electric starting from the Rolls-Royce Derwent, built in Great Britain by the British jet engine pioneer Frank Whittle. As a result, performance was significantly increased and the I-40 version, giving a thrust of 18 kN, was selected to for the Lockeed P-80, the first mass-produced jet fighter aircraft in the USA. Then this led to the two-seat trainer Lockeed T33 (Fig. 3.59), widely used since the 1950s in numerous countries. Given the need of a large number of engines for both aircraft, in 1945 the series production of the J33 engine was entrusted to Allison company, due to its larger capability and costs reduction respect to the General Electric. The J33 was made in different versions, some of which were also conceived for civil aircraft, for a total production of over 6600 units built by Allison and circa 300 units of the first version by General Electric. From a technical point of view, the J33 belongs to the English school turbojets, with a large single-stage centrifugal compressor and an axial-flow turbine. This arrangement, widespread in the 1940s, was quickly abandoned at the beginning of the 1950s in favour of the multi-stage axial-flow compressors (Table 3.29).

Table 3.29 Technical specifications of the Allison J33 turbine

Manufacturer	Allison Engine Company, Indianapolis, USA, from the 1940s up to 1950s
Type	J33
Description	Single-stage axial-flow turbine rotor with 93 blades fitted by jagged connection (Figs. 3.60 and 3.61)
Application	Various military aircraft: Lockheed P-80 *Shooting Star*; Lockheed T-33 *Shooting Star* (trainer version shown in Fig. 3.59); Lockheed F-94A/B Starfire

Fig. 3.59 The two-seater trainer Lockheed T33, powered by a single Allison J33, 1950s

Fig. 3.60 The turbine rotor of a turbojet engine Allison J33, 1945–1950s

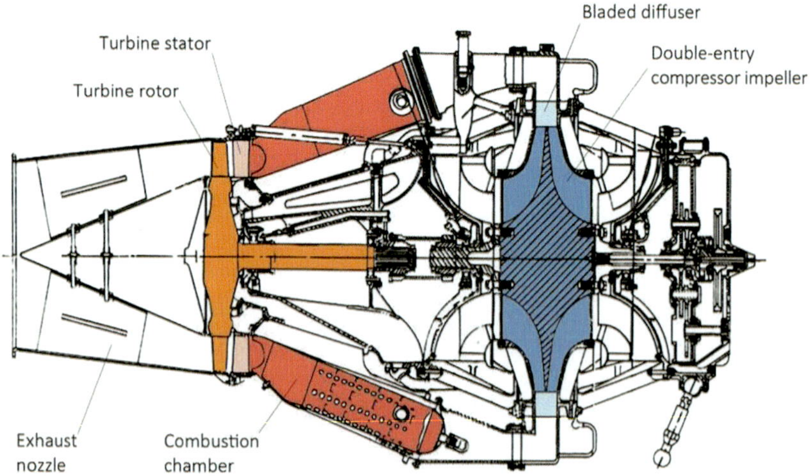

Fig. 3.61 The aircraft turbine wheel Allison J33, 1945–1950s

3.30 General Electric J47

1948–1956

The J47 (Fig. 3.62) was one of first high-performance turbojet engine entirely developed by General Electric from the J35 engine for military aircraft, as well as the first axial-flow turbojet approved for civil aviation in the United States. Respect to the British turbojet engines of that period, it features a multi-stage axial-flow compressor (Table 3.30). The engine was manufactured in various series between 1948 and 1956, even with afterburner, for a total production of over 30,000 units. Over the years it underwent constant development with considerable improvements in terms of performance, reliability, and time between overhauls, which was greatly increased from 15 flight hours in 1948 to 1200 in 1956.

Table 3.30 Technical specifications of the General Electric J47

Manufacturer	General Electric, USA, from 1948 to 1956
Type	J47-GE-13 (Fig. 3.62)
Description	Single-shaft turbojet engine (Fig. 3.63)
Compressor	12-stage axial-flow
Compression ratio	5.0
Airflow	42 kg/s at 7950 rpm at sea level, in static conditions
Combustor	8 direct-flow inter-connected tubular chambers
Turbine	Single-stage axial-flow; stator with 64 internally air-cooled blades; SAE 4340 steel-made impeller core with an outer welded ring made of Timken 16-25-6 alloy with 96 blades inserted by jagged connection; max. turbine inlet gas temperature 870 °C, outlet temperature 690 °C at 7950 rpm
Exhaust nozzle	Fixed geometry cone
Fuel feeding	Double fuel system with two gear pumps Pesco with a barometric control system and additional water injection in the combustion chambers
Lubrication	Forced by one feeding gear pump and 4 scavenge gear pumps, heat exchanger integrated in the storage tank
Max. static thrust	23.14 kN at 7950 rpm, at sea level
Max. continuous thrust	21.03 kN at 7630 rpm, at sea level in static conditions
Specific fuel consumption	1.05 kg/daN/h
Diameter	933 mm
Length	3662 mm
Mass	1134 kg
Mass/Thrust ratio	0.49 kg/daN
Application	Various aircraft: North American F-86 Sabre (Fig. 3.64); North American B-45 Tornado; Martin XB-51; Convair B-36 Peacemaker et al.

Fig. 3.62 The aircraft engine General Electric J47, 1948–1956

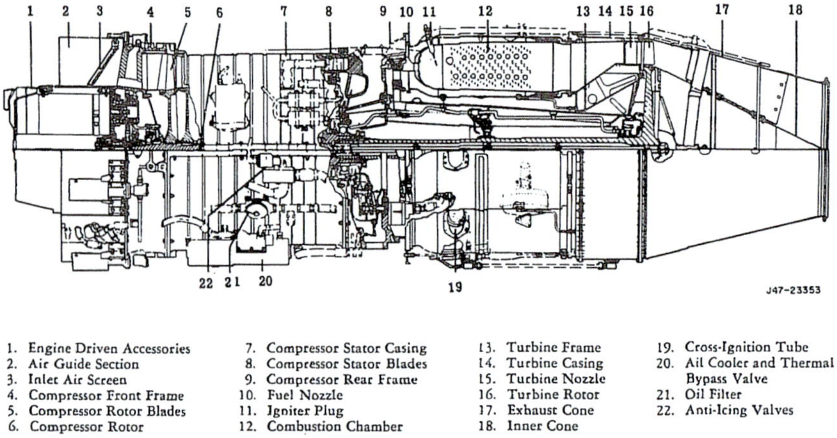

Fig. 3.63 Cross-section view of the aircraft engine General Electric J47

Fig. 3.64 Cross-section view of a North American F-86 Sabre powered by the GE J47

3.31 Turboméca Palas

1950

Turboméca was an important French company specialized in the production of small and medium power turbomachinery for various applications, which today is part of the Safran industrial group. Founded in 1938, the company began its activity with the design of compressors and turbines, followed by the production of compressors and turbochargers for Hispano Suiza aircraft engines. Its first turbojet, named Piméné, was entirely designed in 1941 independently respect to the pioneering studies of Frank Whittle in England and Hans Joachim Pabst von Ohain in Germany. However, due to the Secon World War, the engine was tested only in 1947 and then it was mass produced. Starting from the Piméné, Turboméca developed a new engine, named Palas (Fig. 3.65), using some SNECMA patents and the technical experience gained by eng. Joseph Szydlowski (founder of Turboméca), who collaborated with a group of BMW technicians in the immediate post-war period. The Palas demonstrated excellent performance, thanks to which it was used in many light aircraft and powered gliders of French, English and Italian construction (Table 3.31). It was also used in some traditional propeller-driven aircraft, as auxiliary engine for additional thrust during take-off. Its commercial success was such that it was mass-produced under license by Blackbourn and General Aircraft in the United Kingdom and by the American Teledyne.

Table 3.31 Technical specifications of the Turboméca Palas

Manufacturer	Société Turboméca, Bordes, France, 1950
Type	Palas
Description	Single-shaft turbojet engine
Compressor	Single-stage centrifugal-flow (Fig. 3.66)
Compression ratio	4.12
Max. air flow	3.3 kg/s at 35,000 rpm at sea level
Combustor	Single annular combustion chamber
Turbine	Single-stage axial flow, impeller with 24 blades (25 on later models), max. inlet gas temperature 850 °C
Exhaust nozzle	Fixed geometry cone
Fuel feeding	Rotating disc injector integrated in the crankshaft
Max. static thrust	1.6 kN at 35,000 rpm at sea level
Specific fuel consumption	1.02 kg/daN/h
Diameter	409 mm
Length	1105 mm
Mass	72 kg
Mass/Thrust ratio	0.46 kg/daN
Application	Various aircraft: Caproni Trento F.5; CVV-6 Canguro *Palas*; SIPA S.200 Minijet; Ikarus 451

Fig. 3.65 The aircraft engine Turboméca Palas, 1950. Entrusted by Aeronautica Militare

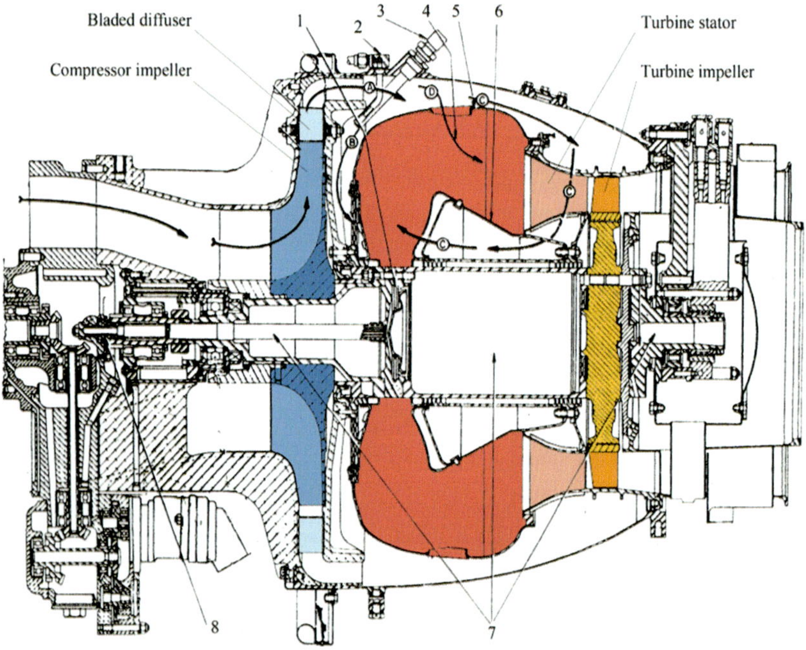

Fig. 3.66 Cross-section view of a Turboméca.turbojet engine. Legend: (1) rotating fuel injector; (2) starting injector; (3) starting spark plug; (4) annular combustion chamber; (5) outer wall of the flame tube; (6) inner wall of the flame tube; (7) crankshaft; (8) fuel inlet at the centre of the crankshaft, with pressure seal element

3.32 FIAT-Bristol Siddeley Orpheus 803

1957–Mid 1960s

The development of the Orpheus turbojet was undertaken by Bristol Siddeley in 1953 and one year later the first version BOr.1 passed the 150-h clearance test giving a thrust of 14.6 kN. As a result, in mid-1955 it was successfully installed in the Folland Gnat, a light fighter prototype. In 1956 some engines of the BOr.2 version were subjected to very demanding tests with atmospheric conditions of +43 °C and 80% air humidity and then in an almost arctic climate with temperatures of −34 °C, demonstrating excellent reliability even in extreme operating conditions. Given the excellent results, in 1957 the engine was further developed in the BOr.3 version (Fig. 3.67), which was chosen to power the new FIAT G. 91 light fighter, in active service for 30 years in various NATO countries. The engine achieved considerable success and was also produced under license in India, France, Germany and Italy, where FIAT built the 803 version (BOr.3). Specifically designed for light fighter aircraft, the Orpheus features a high thrust-to-mass ratio, the best among comparable engines of its category in 1957 (version 803/BOr.3). The front part is made of light alloy while the body of the compressor and all the accessories are grouped in a single-piece case made of magnesium and zirconium alloy. The engine shaft is supported by only two bearings located at the front and rear ends, i.e., respectively before the compressor wheels and downstream of the single-stage turbine (Table 3.32).

Table 3.32 Technical specifications of the FIAT-Bristol Siddeley Orpheus 803

Manufacturer	FIAT, Turin, from 1957 until the mid-1960s, under license from Bristol Siddeley Engines Ltd., UK
Type	Orpheus 803 (FIAT model 4023.002)
Description	Single shaft turbojet engine
Compressor	7 stages axial-flow (Fig. 3.68)
Compression ratio	4.4
Max. air flow	38 kg/s at 10,000 rpm at sea level, in static conditions
Combustor	7 tubular combustion chambers
Turbine	Single-stage axial-flow; stator (nozzle) with 63 blades; impeller with 125 blades
Exhaust nozzle	Fixed geometry cone
Fuel feeding	Variable stroke piston pump with Lucas automatic control system for fuel flow and air/fuel ratio
Lubrication	Forced by feeding and scavenge pumps
Max. static thrust	22.3 kN at 10,000 rpm
Specific fuel consumption	1.10 kg/daN/h
Diameter	823 mm
Length	2441 mm
Mass	475 kg
Mass/Thrust ratio	0.21 kg/daN
Application	Light fighter aircraft FIAT G.91 R and T (Fig. 3.69)

Fig. 3.67 The aircraft engine FIAT-Bristol Siddeley Orpheus 803, 1957–mid 1960s. Entrusted by Aeronautica Militare

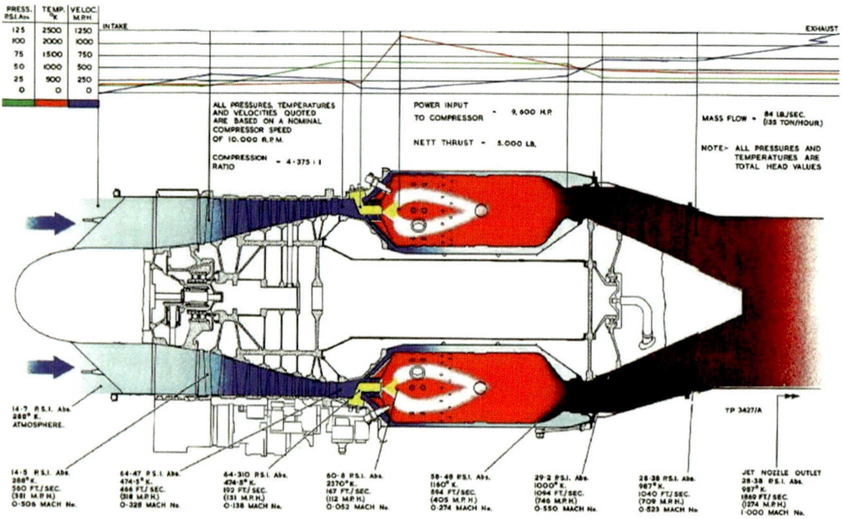

Fig. 3.68 Schematic cross-section view of the turbojet engine Bristol Siddeley Orpheus 803, with its working cycle

Fig. 3.69 Cross-section view of the FIAT G.91 showing the turbojet engine Bristol Siddeley Orpheus 803 located at the rear part of the aircraft with a long internal air inlet duct

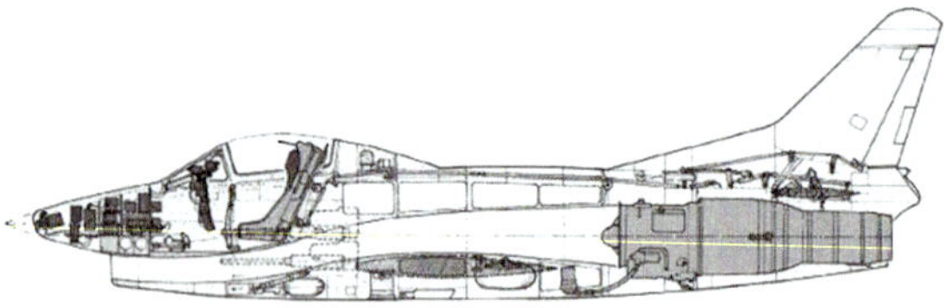

3.33 FIAT-General Electric J79-GE-19

1960s–1970s

The General Electric J79 was one of the most successful high-performance jet engines in aviation history, whose project dates to the early 1950s, since it was originally designed for the supersonic interceptor Lockheed F-104. Since the first prototypes, the engine featured many innovative characteristics of high value. Later, numerous models with several technical and performance improvements were developed for military and civil application. The J79 had a very long operational life, with a mass production lasting over 30 years: in total more than 17,000 units were built in the United States and in several other countries under license. In Italy it was produced in various versions by FIAT and Alfa Romeo for the F-104 of the Italian Air Force (Table 3.33).

Table 3.33 Technical specifications of the FIAT-General Electric J79-GE-19

Manufacturer	FIAT, Turin, from the mid-60s to the 70s, under license of General Electric (USA)
Type	J79-GE-19 (Fig. 3.70)
Description	Single-shaft turbojet engine for supersonic aircraft
Compressor	17 stages axial-flow, with variable pitch stator blades at inlet and following first 6 stages (Fig. 3.71)
Compression ratio	12.7
Air mass flow	77 kg/s at 7460 rpm
Combustor	10 can-type combustion chambers and afterburner downstream the turbine
Turbine	3 stages axial-flow, max. inlet gas temperature 990 °C
Exhaust nozzle	Variable geometry nozzle, max. temperature of outgoing gas 1980 °C in case of operative afterburner
Fuel feeding	Automatic fuel flow regulator, including afterburner fuel nozzles control
Lubrication system	Forced by feeding and scavenge pumps, with two heat exchangers
Max. static thrust	79.6 kN (with afterburner)
Max. continuous thrust	52.8 kN
Specific fuel consumption	1.97 kg/daN/h
Diameter	998 mm
Length	5334 mm
Mass	1740 kg including afterburner
Max. mass/Thrust ratio	0.22 kg/daN
Application	Fighter aircraft Lockheed F-104 S and following upgraded models (Fig. 3.72)

Fig. 3.70 The aircraft engine FIAT-General Electric J79-GE-19, 1960s–1970s. Entrusted by Aeronautica Militare

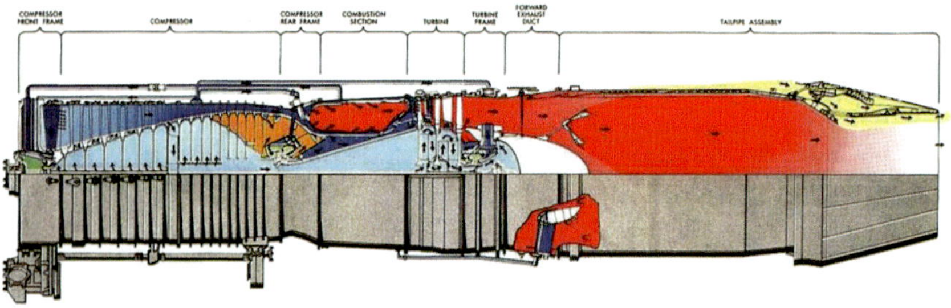

Fig. 3.71 The aircraft engine FIAT-General Electric J79-GE-19, 1960s–1970s

Fig. 3.72 The aircraft engine FIAT-General Electric J79-GE-19, 1960s–1970s

3.34 Armstrong Siddeley Viper 22

Late 1950s–1970s

At the beginning of the 1950s the British company Armstrong Siddeley designed a turbojet engine, named Viper, for light aircraft and trainers. Its production lasted for over twenty-five years, even when the original manufacturer merged with the Bristol company in 1960 and when the latter was in turn acquired by the Rolls Royce Group in 1966. Initially conceived as an expendable short-life engine for target drone, the Viper was developed in several versions for many crewed aircraft, with continuous technical upgrades and performance improvements, ranging from 6.8 kN thrust of the first models to over 19 kN in the latest ones. The engine features a common axial-flow arrangement and an annular combustion chamber with multiples concentric flame tubes, each one equipped with a particular fuel-vaporizing injector (Table 3.34). This system was designed and developed by Sid Allen at Armstrong Siddeley and later was used by other manufacturers. Thanks to its performance and versatility, the engine achieved considerable commercial success, both in military and civil application. The Viper was built under license in various countries, such as Romania, ex Jugoslavia and Italy, where Piaggio produced the mk. 22 model for the aircraft trainer Aermacchi MB-326, while FIAT the more powerful series 600.

Table 3.34 Technical specifications of the Armstrong Siddeley Viper 22

Manufacturer	Armstrong Siddeley, UK, mid 1950s to late 1970s
Type	Viper 22 (Fig. 3.73)
Description	Single-shaft turbojet engine (Fig. 3.74)
Compressor	7 stages axial flow
Compression ration	4.1
Air mass flow	14.5 kg/s at 13,800 rpm at sea level
Combustor	Annular chamber with 24 concentric flame tubes
Turbine	Single-stage axial flow
Exhaust nozzle	Fixed geometry cone
Fuel feeding	One vaporizing fuel injector (Armstrong Siddeley system) in each flame tube
Lubrication	Forced by gear pumps
Max. thrust static	13.3 kN at 13,800 rpm at sea level
Specific fuel consumption	1.0 kg/daN/h
Diameter	617 mm
Length	1703 mm
Mass	284 kg
Max. mass/Thrust ratio	0.21 kg/daN
Application (mk. 22 model)	Aircraft trainer Aermacchi MB-326

Fig. 3.73 The aircraft engine Armstrong Siddeley Viper 22, late 1950s–1970s. Entrusted by Aeronautica Militare

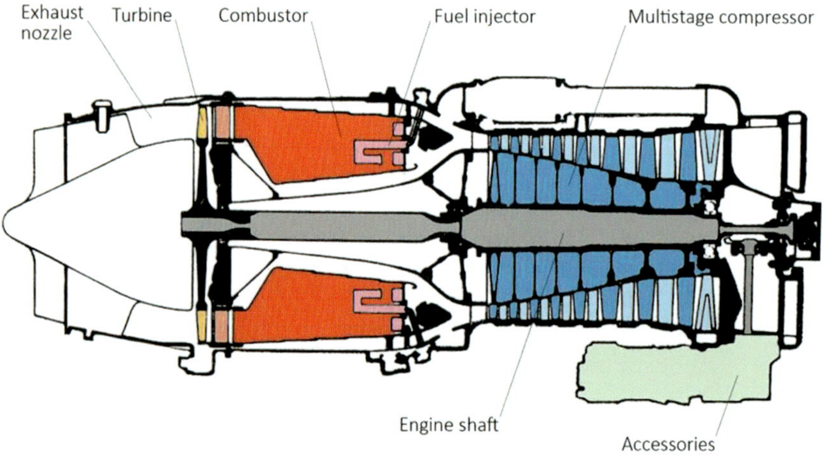

Fig. 3.74 Schematic cross-section view of an aircraft engine Armstrong Siddeley Viper

3.35 Bristol Siddeley Gnome H.1000

1960s

The Gnome is a turboshaft engine with a free power turbine, that was strictly derived from the General Electric T58 engine, whose de Havilland acquired the production license in the late 1950s. The Gnome production, as well as its development program, was maintained when de Havilland was acquired by Bristol Siddeley and, later, when the latter was in turn acquired by Rolls-Royce. The Gnome engine was made in two versions: the turboshaft model, named H-series (Fig. 3.75), specifically designed for helicopters, and the turboprop model, P-series, used for airplanes. Both were produced in various subtypes, both for military and civil applications. The Gnome engine was one of the first jet engines to be equipped with an (analogue) electronic control system of the fuel feed for simplifying the engine load management by the pilot and increasing the safety against accidental overloads (Table 3.35). This system was designed specifically for the Gnome engine in Great Britain in collaboration with Lucas company and it can be considered as the forerunner of all subsequent FADEC (Full Authority Digital Engine Control) systems.

Table 3.35 Technical specifications of the Bristol Siddeley Gnome H.1000

Manufacturer	Bristol Siddeley (later Rolls-Royce), UK, 1960s
Type	Gnome, H.1000 version
Description	Turboshaft engine with free power turbine
Compressor	10 stages axial flow, with variable pitch stator blades at the first 3 stages
Compression ratio	8.3
Max. air flow	Approximately 6.0 kg/s at 26,300 rpm at sea level
Combustor	Annular combustion chamber
Gas generator turbine	2 stages axial-flow, max. inlet temperature 595 °C
Power turbine	Single stage axial-flow
Fuel feeding	Lucas electronic control system with hydraulic actuators
Lubrication	Forced by gear pumps
Max. power	972 hp at 19,300 rpm at sea level (short time/take-off)
Specific consumption	Circa 286 g/kW/h
Primary drive ratio	2.926 by means of a speed reducer with herringbone gears
Diameter	406 mm
Length	1676 mm
Mass	210 kg (including accessories, primary reduction gear and related drive shaft)
Mass-to-power ratio	0.22 kg/hp
Application	Various helicopter: Westland Sea King, Wessex and Whirlwind (Fig. 3.76); Kawasaki KV-107 (licensed model of Boeing CH-46); Agusta A.101

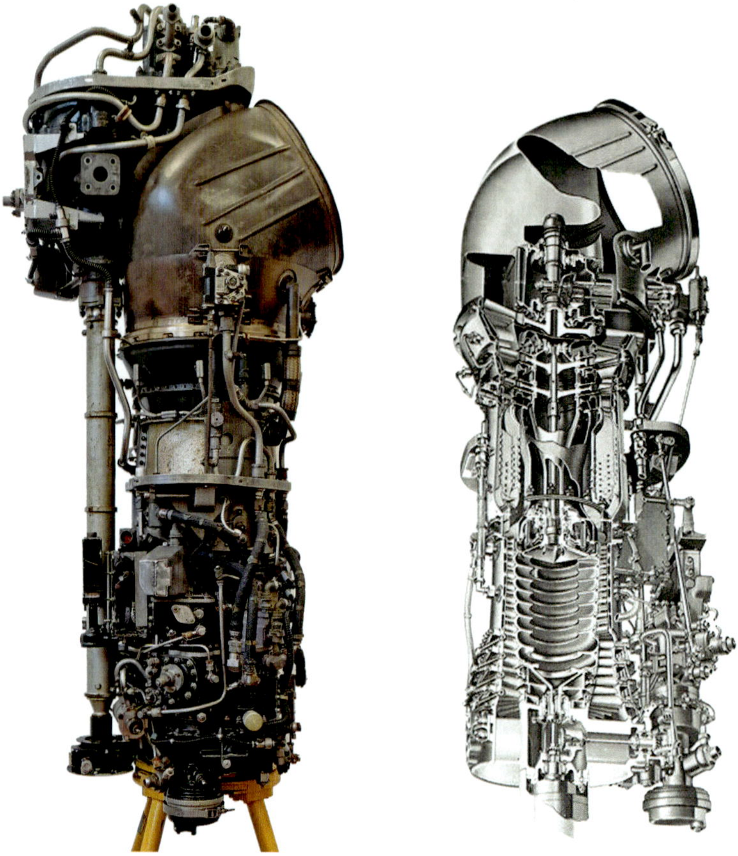

Fig. 3.75 The Bristol Siddeley Gnome turboshaft (left) with its cross-section view (right). Entrusted by Aeronautica Militare

Fig. 3.76 A Westland Whirlwind helicopter with a Bristol Siddeley Gnome turboshaft engine

3.36 Rolls-Royce RB.162-81

1961–1970

The Rolls-Royce RB.162 is a jet engine specifically designed to provide vertical thrust in VTOL (Vertical Take-Off and Landing) aircraft. Its development program, based on various tests carried out in the first half of the 1950s, commenced in 1959 through a collaboration agreement between the governments of the United Kingdom, France and West Germany. Derived from the 1955 RB.108 engine, the prototype of the RB.162 was completed at the end of 1961 and then different models were built for tests. Numerous experimental flights were carried out by means of the prototypes of three new VTOL aircraft: the Dornier Do 31, the Dassault Mirage IIIV, and the VFW VAK 191B, characterized by different propulsion systems and intended uses. To obtain the highest levels of lightness, simplicity and reliability, the designers experienced many innovative technical features, including reinforced composite materials and a remarkable simplification of the engine layout, as well as, of the auxiliary systems (Table 3.36). Despite positive results, no aircraft achieved a mass-production but the RB.162 engine, built in about 86 units, paved the way for the use of composite materials in aircraft engines.

Table 3.36 Technical specifications of the Rolls-Royce RB.162-81

Manufacturer	Rolls-Royce, Derby, UK, from 1961 to 1970
Type	RB.162-81 (Fig. 3.77)
Description	Single-shaft turbojet engine for vertical take-off and landing aircraft
Compressor	6 stages axial flow, stator and blades made of fibreglass reinforced epoxy resin, except the first stage blades made of aluminium alloy
Compression ratio	4.25
Air mass flow	41.3 kg/s
Combustor	Annular combustion chamber
Turbine	Single-stage axial flow, with a titanium alloy impeller and air-cooled blades made of Nimonic (a nichel- chromium-based alloy)
Exhaust nozzle	Fixed geometry cone
Lubrication	Open loop system, with disposable in-pressure oil sent to the two crankshaft bearings, consumption 1.4 L/h
Starting system	Pneumatic (high pressure air to turbine impeller)
Max. static thrust	26.7 kN
Mass	127 kg
Mass/Thrust ratio	0.05 kg/daN
Application	Various VTOL aircraft: VFW VAK 191B (model 81, Fig. 3.78); Dornier Do 31; Dassault Mirage IIIV

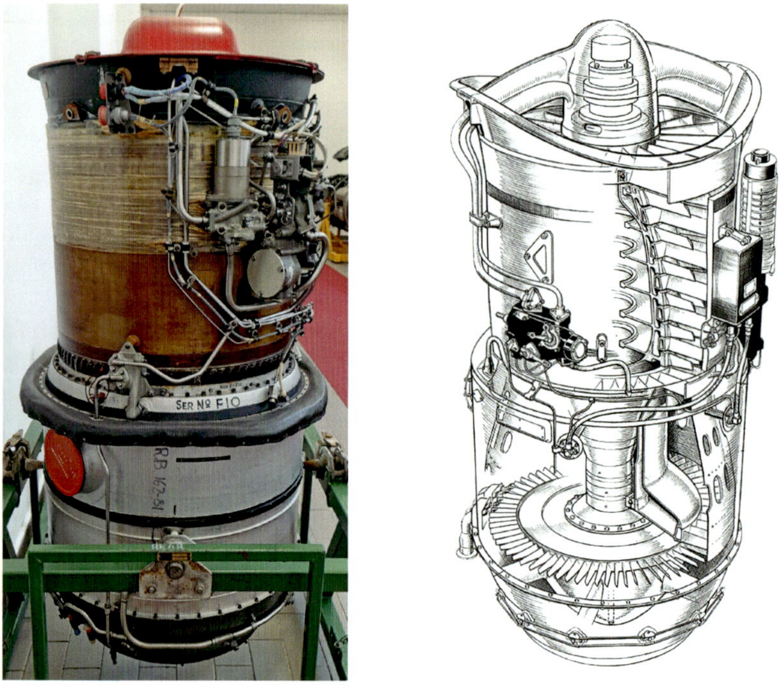

Fig. 3.77 Rolls-Royce RB.162 turbojet lifter engine (left) with its cross-section view (right). Entrusted by Aeronautica Militare

Fig. 3.78 The VTOL fighter aircraft prototype VFW VAK 191B powered by a Rolls-Royce/MAN Turbo RB193-12 vectored-thrust turbofan engine for lift/cruise and two (front and rear) Rolls-Royce RB.162 as lifter engines for vertical take-off and landing, early 1970s

3.37 Fichtel & Sachs KM 48

1965–1975

In the early 1950s the invertor Felix Wankel designed a new and innovative rotary engine which shortly found a large variety of practical applications. The first prototype was completed in 1957 at the NSU company and then many other industries bought the manufacturing license. The Fichtel & Sachs started to produce Wankel-type engines at the beginning of the 1960s for stationary equipment, boats, motorcycles and small agricultural machinery. One of the first model was the 1965 KM 48, initially conceived as a stationary engine and later adapted for various applications including motor gliders (Fig. 3.79), with the K8B version. Thanks to its compact size and good mass-to-power ratio, it became one of the first Wankel engines used in aviation. Its mass production was later concluded in 1975 due to the oil crisis (Table 3.37).

Table 3.37 Technical specifications of the Fichtel & Sachs KM 48

Manufacturer	Fichtel & Sachs AG, Schweinfurt, Germany, from 1965 to 1975
Type	KM 48 (Fig. 3.80)
Description	Wankel rotary engine (Fig. 3.81), single rotor, crankcase made of aluminium alloy
Displacement	160 cm^3
Compression ratio	8.0
Power	8 hp at 4700 rpm 10 CV at 5000 rpm from K8B model
Fuel specific consumption	315 g/hp/h
Specific power	50.0 hp/litre
Fuel feeding	Single barrel petrol carburettor
Ignition	Single spark plug supplied by an inductive discharge high voltage distributor, powered by a dynamo
Cooling	Air-fed forced by a crankshaft-driven fan
Lubrication	50:1 petrol-oil mixture
Application	Motorboats, motorcycles, small agricultural machines, motor gliders (for auxiliary thrust, not for take-off)

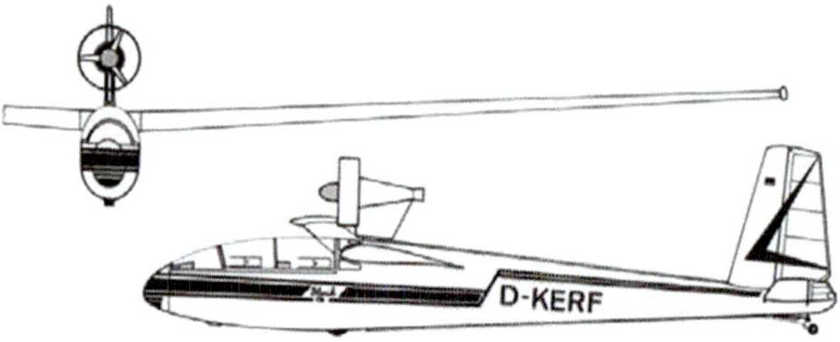

Fig. 3.79 The Blanik motor glider with an auxiliary Fichtel and Sachs Wankel engine, 1974

Fig. 3.80 The aircraft engine Fichtel & Sachs KM 48, 1965–1975

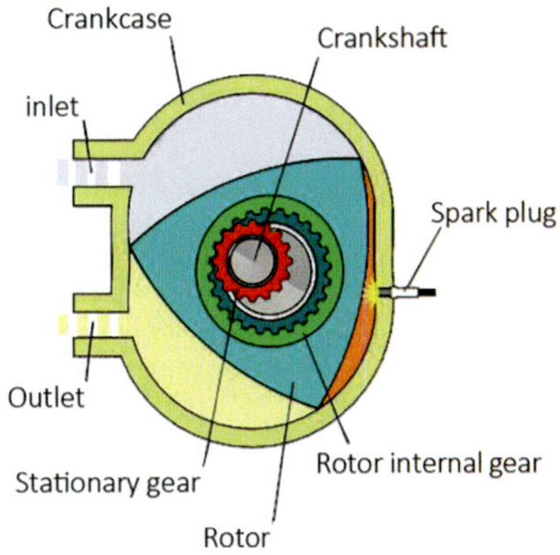

Fig. 3.81 Schematic cross-section view of a Wankel rotary engine

3.38 Moto Guzzi V 50 Aircraft Engine Prototype

1970s

Between the end of the 1960s and the beginning of the 1970s, Moto Guzzi received some requests to supply an engine for unmanned and ultralight aircraft, on the basis of which carried out some studies, with the design of some prototypes. The engine on display is a mock-up prototype, created during that development process strictly derived from the V50 motorcycle engine (Figs. 3.82 and 3.83). Compared to the series version, the prototype features a propeller speed reducer and a specific horizontal single-barrel carburettor. Furthermore, the exhaust system is completely different from the motorcycle model, since its layout is conceived to favour both inlet air and then air–fuel mixture heating by means of heat exchange with the exhaust gas, to improve the carburetion process at low temperatures at high altitude (Table 3.38).

Table 3.38 Technical specifications of the Moto Guzzi V 50 aircraft engine prototype

Manufacturer	Moto Guzzi, Mandello del Lario (Lecco), Italy, 1970s
Type	Mock-up prototype derived from the Moto Guzzi V50 motorcycle engine (Fig. 3.84)
Description	Two-cylinder V-90° engine, crankcase, cylinders, and heads made of aluminium alloy
Stroke	57 mm
Bore	74 mm
Displacement	490 cm^3
Compression ratio	10.5
Power	47 hp at 7500 rpm
Valvetrain	2 overhead valves per cylinder, operated by pushrods and rocker arms by a single camshaft in the crankcase
Fuel feeding	One single-barrel petrol carburettor Dell'Orto type OVC-40-37-A
Ignition	Single spark plug per cylinder
Cooling	Air-fed
Lubrication	Forced by a pump
Propeller transmission	By means of a spur gears reducer
Application	Unmanned and ultralight aircraft

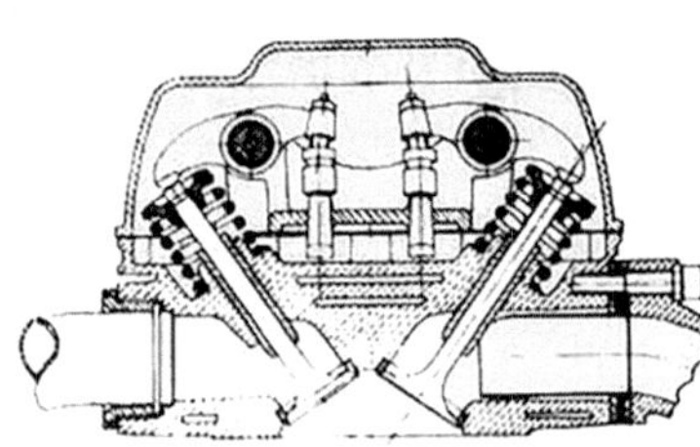

Fig. 3.82 Cross-section view of the engine head of a Moto Guzzi V 50

Fig. 3.83 The aircraft engine prototype derived from the Moto Guzzi V50 motorcycle, 1970s. Kind donation of Mr. Girolamo Di Giovanni

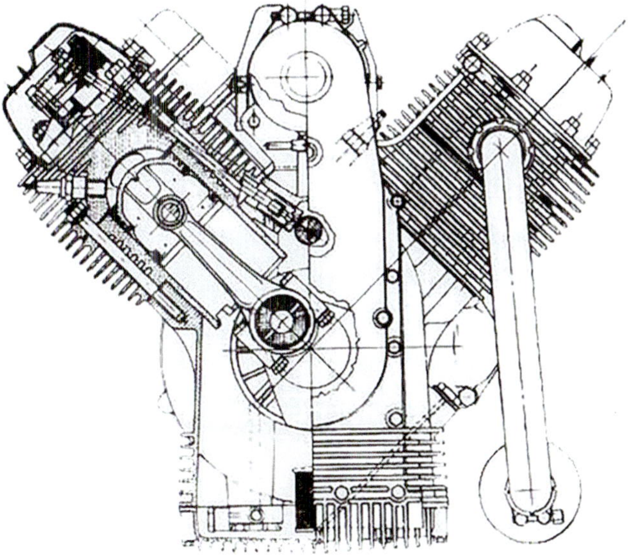

Fig. 3.84 Cross-section view of the motorcycle engine Moto Guzzi V 50, 1970s

Chapter 4
Aeroplanes

Abstract The Museum has two historic aircraft on display: a FIAT G. 59 4B trainer and an Aeritalia–Lockheed F-104S ASA-M. The first represents one of the last high-performance piston engine military aircraft to be produced at the end of 1940s. Originally conceived and used as advanced trainer aircraft, it was purchased by University of Palermo in late 1960s for educational purpose. The second, instead, represents the new modern era of aviation since it is one of the first supersonic mass-produced military aircraft. It was used in numerous countries, and, thanks to the development of many upgraded versions, some latest models remained in service up to 2004, after fifty years from its first flight.

G. Genchi, *The Museum of Engines and Mechanisms of the University of Palermo*,
History of Mechanism and Machine Science 48,
https://doi.org/10.1007/978-3-031-63042-2_4

4.1 FIAT G. 59 4 B

1952

The FIAT G. 59 was one of the last high-performance aircraft equipped with a piston engine and it can be also considered as a symbol of the post-war re-birth of the Italian aeronautical industry. It was designed by famous Italian aeronautical engineers Giuseppe Gabrielli, starting from the FIAT G. 55 Centauro, considered one of the best Italian fighter aircraft of the Second World War. The G. 59 was produced by FIAT in the early 1950s, in more than 180 copies mainly used in the Italian Air Force as advanced trainer aircraft. Thanks to its engine, a 1660 hp V-12-cylinder Rolls Royce Merlin (version 500–20), the aircraft could reach a top speed of 609 km/h at an altitude of 6400 m above the sea level, and a ceiling of 12,100 m. The Museum's FIAT G. 59 4B is a two-seater model (Figs. 4.1 and 4.2) which was in service at the Italian Air Force up to 1964. Then it was purchased for didactic purpose by the former Institute of Aeronautics of the University of Palermo and in 2013 it undergone a thorough restoration in the Museum's workshop. It is one of the only five complete G. 59 surviving today, and it is displayed in a special area of the Museum (Table 4.1).

Table 4.1 Technical specifications of the FIAT G. 59 4 B

Manufacturer	FIAT, Turin, Italy, 1952
Type	G. 59 4B
Series	VII, identification codes MM 53530, N.C. 115, RR 73
Description	Two-seater (tandem) dual-controls training aircraft with all-metal airframe (Figs. 4.2 and 4.3)
Max. speed	609 km/h at 6400 m above the sea level
Max. operational ceiling	12,100 m
Max. Rate of climb	64 km/h at 2000 m
Mass	2850 kg (dry)
Wing area	21.11 m^2
Wingspan	11,860 mm
Length	9476 mm
Engine	Rolls-Royce Merlin 500–20, V-60° 12-cylinder engine, liquid-cooled, with a centrifugal compressor driven by the engine by means of a 2-speed transmission (Fig. 4.4)
Power	1660 hp at 3000 rpm
Propeller	All-metal four-blade variable pitch propeller FIAT Hamilton 5010
Fuel-capacity	350 L in normal condition and 600 L with two external tanks
Flight range	762 km at 488 km/h at 5600 m, 1352 km at 460 km/h at 5500 m with two external tanks

Fig. 4.1 The trainer aircraft FIAT G. 59 4B of the Museum (MM 53530—RR 73), 1952

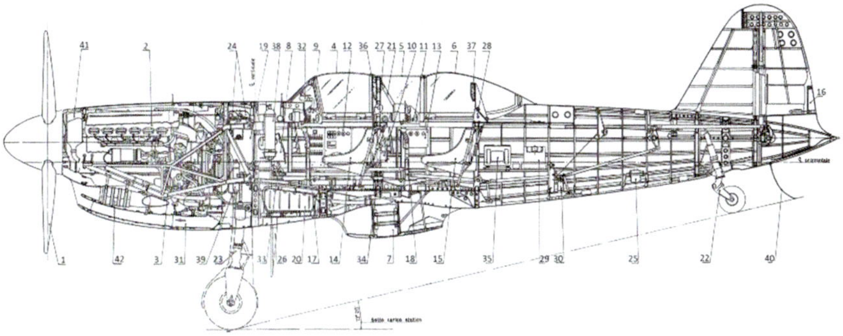

Fig. 4.2 Cross-section view of the trainer aircraft FIAT G. 59 4B

Fig. 4.3 The front cockpit of the trainer aircraft FIAT G. 59 4B

The Rolls-Royce Merlin is considered one of the best piston aero engines of the 1940s. Its original project dates to the early 1930s and was based on the experiences acquired by Rolls-Royce through various high-performance 12-cylinder engines, including the "R" competition model used with success in the Schneider Trophy. The Merlin was produced in many series and types, characterized by continuous technical and performance improvements. During the Second World War it was used in numerous military aircraft, including the famous Supermarine Spitfire. Developed after the war, the Merlin 500 series (Tab. 4.2) features a two-speed single-stage supercharger (Fig. 4.4) and was used in Italy to equip the FIAT G. 59 (Fig. 4.5 and Table 4.2).

Table 4.2 Technical specifications of the Rolls-Royce Merlin 500–20

Manufacturer	Rolls-Royce Limited, UK, from 1945 to early 1950s
Type	Merlin 550–20 (derived from T24 model)
Description	V-60° 12-cylinder engine, crankcase, heads, and pistons made of aluminium-alloy (Fig. 4.6)
Stroke	152.4 mm
Bore	137.2 mm
Displacement	27.0 L
Compression ratio	6.0
Power	1660 hp at 3000 rpm at 690 m above sea level (5 min.) 1532 hp at 3000 rpm at 2800 m above sea level
Valvetrain	4 overhead valves per cylinder, controlled via rocker arms by two overhead camshafts in each bank
Fuel feeding	Single twin-barrel petrol carburettor, placed upstream of the compressor
Supercharging	Single-stage centrifugal compressor, crankshaft driven by an automatic 2-speed gearbox (Fig. 4.4)
Ignition	Two spark plugs per cylinder supplied by two independent high-voltage magnetos
Cooling	Liquid-fed by a centrifugal pump
Lubrication	Forced by feeding and scavenge gear pumps
Mass	658 kg (dry)
Mass-to-power ratio	0.40 kg/hp
Propeller speed ratio	0.420 by means of a spur gears transmission

Fig. 4.4 Schematic layout of the compressor two-speed gear train driven by the crankshaft

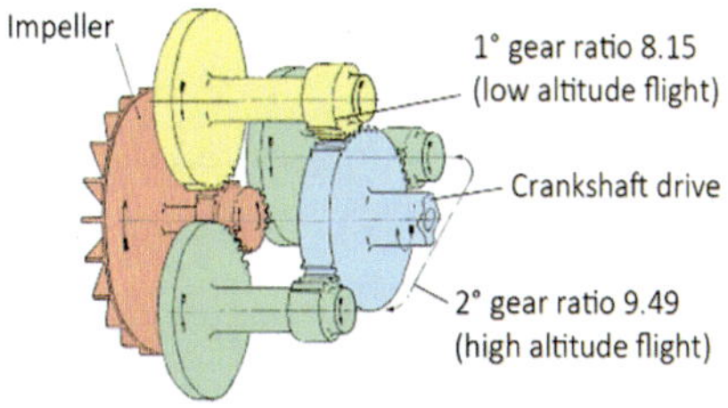

Fig. 4.5 Close view of the FIAT G. 59 4B showing its 12-cylinder engine Rolls-Royce 550–20

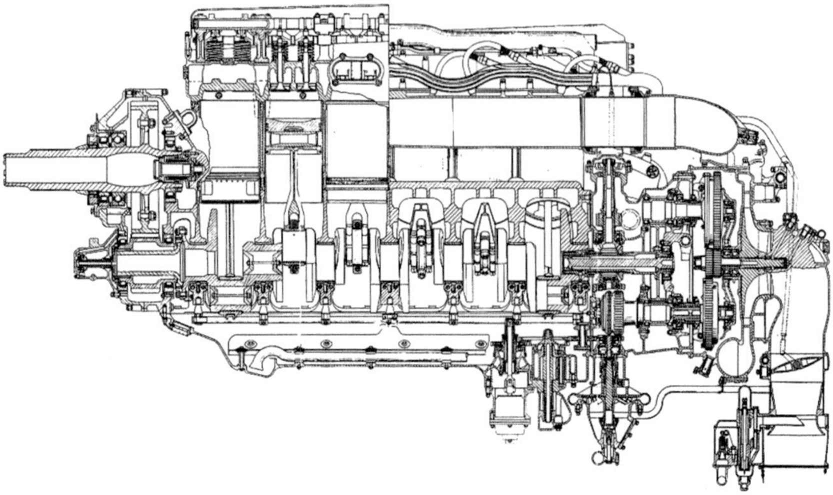

Fig. 4.6 Cross-section view of the 12-cylinder engine Rolls-Royce 550–20

4.2 Aeritalia-Lockheed F-104S ASA-M

1979

The Lockheed F-104 Starfighter is a supersonic all-weather interceptor aircraft, developed in the early 1950s in response to the rapid increase in performance of Soviet jet aircraft, in particular the Mig-15. In two years, the designers and technicians led by Eng. Clarence "Kelly" Johnson managed to complete the first prototype, that was tested in 1954. The first F-104s became operational from 1958 in various wings of the United States Air Force (USAF). Produced and developed in many construction series, characterized by multiple improvements in performance and operational capabilities, the F-104s were used for a long time in various countries. In Italy, the F-104s were built under license by Aeritalia, in different series with continuous updates, for a total production of slightly more than 200 units up to 1979. They were used in various wings of the Italian Air Force until 2004. From a technical point of view, the aircraft features a particular structure with a long fuselage, short half wings, a raised T-shaped tail, and the use of a single General Electric J79 engine. Specifically designed for the F-104, the J79 was an innovative high-performance turbojet engine, with advanced features such as variable stator compressor, afterburner, and variable-geometry exhaust nozzle. This has long been one of the most technically advanced engines, capable of propelling the aircraft over double the speed of sound (2.2 Mach). The F-104 on display (Fig. 4.7) belongs to the S version, the last one produced until the end of the 70s, it and was subject to various modernizations up to the last operational configuration ASA-M, introduced in 1997. Withdrawn from operational service in 2004, it has been placed on display at the academic campus of the University of Palermo (Table 4.3).

Table 4.3 Technical specifications of the Aeritalia—Lockheed F-104S ASA-M

Manufacturer	Aeritalia under license of Lockheed, Turin, Italy, 246 aircraft built from 1966 up to March 1979
Type	F-104S ASA-M, identification code MM 6714
Description	supersonic single-seat aircraft (Fig. 4.8)
Engine	Single turbojet General Electric J79-GE-19 (Tab. 3.33)
Wing area	18.2 m^2
Wingspan	7.6 m
Length	16.7 m
Height	4.1 m
Empty mass	6760 kg (empty), 9840 kg (gross)
Fuel capacity	3,392 L in 5 fuselage bag tanks, with provision for 2 × 740 L under-wing drop tanks and 2 × 645 L wing-tip tanks
Max. speed	2,330 km/h, i.e. Mach 2.2 at 11,000 m above seal level
Max. rate of climb	277 m/s, from sea level to 17,100 m in 2 min. 40 s
Service ceiling	18,000 m (depending on configuration/payload)
Flight range	2920 km (ferry) with additional external tanks

Fig. 4.7 The Aeritalia-Lockheed F-104S ASA-M on display at the academic campus of the University of Palermo. Entrusted by Aeronautica Militare

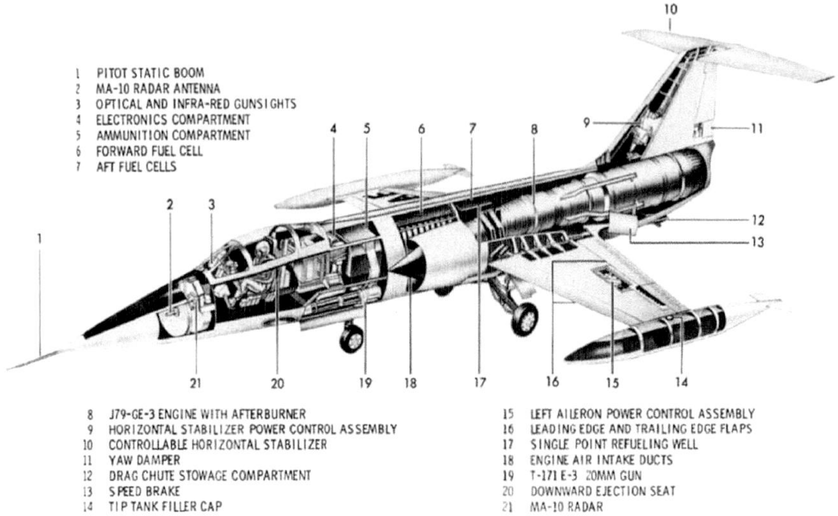

Fig. 4.8 Transparent view of a F-104 showing its main components

Chapter 5
Amotive and Motorcycle Engines

Abstract The automotive section of the museum displays a heterogeneous range of engines spanning over a century, from the 1919 FIAT type 101 engine to the modern Toyota 2ZR hybrid powertrain. It encompasses various engineering solutions, materials, and construction techniques evolving from semi-artisanal to modern industrial production. The transition from carburettors to fuel injection systems, including electronic controls, reflects advancements aimed at enhancing engine efficiency and reducing emissions. Comparisons between displayed engines highlight many technological progresses, such as on considering the numerous differences between the 1929 8-cylinder 7.4 litres Isotta Fraschini 8A producing the same 115 hp power as the 2009 Ford 3-cylinder 1.0 litre EcoBoost. Some engines are exhibited with gearboxes, showcasing different drivetrain arrangements, from the archetypal sliding mesh gearbox up to advanced dual-split continuous variable transmission, passing through the fine transaxle layout of the Lancia Flaminia 2500. The collection includes rare and valuable engines that of the FIAT 8V, the FIAT-Ferrari Dinos up to the modern Ferrari 430 and Lamborghini Aventador. The same section features motorcycle and truck engines, including specimen of fine design like the Lancia-Junkers Tipo 89 and experimental prototypes.

G. Genchi, *The Museum of Engines and Mechanisms of the University of Palermo*, History of Mechanism and Machine Science 48,
https://doi.org/10.1007/978-3-031-63042-2_5

Fig. 5.1 The exhibit area of the Museum dedicated to the automotive and motorcycle engines

The automotive engine section (Fig. 5.1) is made up of various examples which together represent a span of more than a century, from the FIAT type 101 engine from the first post-war period to the modern Toyota 2ZR hybrid engine. Therefore, all the main engineering solutions that have followed one another over time are well represented in it, as regards the structure of the engines, the materials used and the construction techniques, from the semi-artisanal production still in use in the 1920s up to the modern industrial production with numerous automated processes. The same thing applies to the ancillary devices, including first and foremost the carburettor fuel systems, progressively replaced by fuel injection systems, initially mechanical and then those with electronic control, integrated with the ignition system in petrol and petrol-powered engines gas. The large-scale diffusion of electronic systems is closely connected to the need to increase engine efficiency and contain polluting emissions, through the use of exhaust gas treatment devices. Often the comparison between some examples on display allows us to clearly illustrate the results of the technological progress achieved in automotive: from the prestigious 1929 7.4 litres 8-cylinder engine Isotta Fraschini 8A, giving the same 115 hp max. power output of the 2009 1.0 litre 3-cylinder Ford EcoBoost. Some engines are exhibited with their own gearbox showing various drivetrain arrangement. The most interesting of these is the engine of the Lancia Flaminia 2500, with the entire clutch-gearbox-differential unit in a single block according to the transaxle layout. Similarly to the others already mentioned, this part of the Museum's collection was built up over time starting from various engines gradually acquired for experimental research activities and then progressively fallen into disuse due to their obsolescence. Often some have been modified several times to adapt to the experimental activities and, therefore, during the restoration it was necessary to restore their originality. Other specimens, on the other hand, were acquired for didactic activities or, for various reasons, have never been used, remaining in absolutely original conditions, with a high philological and collector's value. One of these is the rare FIAT 8V engine, made for the sports car of the same name, produced by FIAT in only 112 units, it was successfully used in numerous sporting competitions. Sports engines also include the FIAT Dino, 2000 and 2400 engines, the result of a collaboration between FIAT and Ferrari, the Lancia Turbo 1.6 up to the contemporary 8-cylinder Ferrari F136 and 12-cylinder Lamborghini *Aventador* L539. In the automotive engine section, motorcycle and heavy vehicle engines are also currently on display. Among these, the rarest and most technically refined is the Lancia-Junkers Tipo 89, a two-stroke Diesel engine from the 1930s, characterized by the use of two opposing pistons inside each cylinder. The same principle, albeit implemented with a different crank mechanism, is present in the equally rare 2 H 88 engine, built by Breda in the late 1940s for agricultural tractor. The collection also includes a Wankel engine and an experimental prototype of a rotary engine with toroidal combustion chambers, built at the Machinery Institute of the University of Palermo in the 1960s.

5.1 FIAT 101

1919–1926

Designed at the end of the First World War to equip the FIAT 501 car (Fig. 5.2), the type 101 engine was produced in large series from 1919 to 1925; it was also used for the FIAT 502, a long wheelbase chassis version of the 501, until 1926. The engine overall construction arrangement is simple and rational, with quite common features at the time: the four cylinders are grouped in a single cast-iron block mounted over an aluminium alloy crankcase, and the combustion chamber is a flat-shape type with side valves. The engines of the second series were endowed with a removable head

Fig. 5.2 A FIAT 501 powered by the engine FIAT 101

Table 5.1 Technical specifications of the FIAT 101

Manufacturer	FIAT, Turin, Italy, from 1919 to 1926
Type	101
Description	Inline 4-cylinder engine, aluminium alloy crankcase and cast-iron cylinder block (Fig. 5.4)
Stroke	110 mm
Bore	65 mm
Displacement	1460 cm^3
Compression ratio	4.3
Power	23 hp at 2600 rpm
Specific power	15.8 hp/litre
Valvetrain	2 side valves per cylinder, controlled by a camshaft housed in the crankcase, driven by a pair of gears
Fuel feeding	Single barrel petrol carburettor
Ignition	Single spark plug per cylinder supplied by a high-voltage magneto Marelli MP4
Cooling	Liquid-fed by a centrifugal pump
Lubrication	Forced by a gear pump
Application	FIAT 501, FIAT 502 and FIAT 503 (27 hp)

Fig. 5.3 The automotive engine FIAT 101, 1919–1926

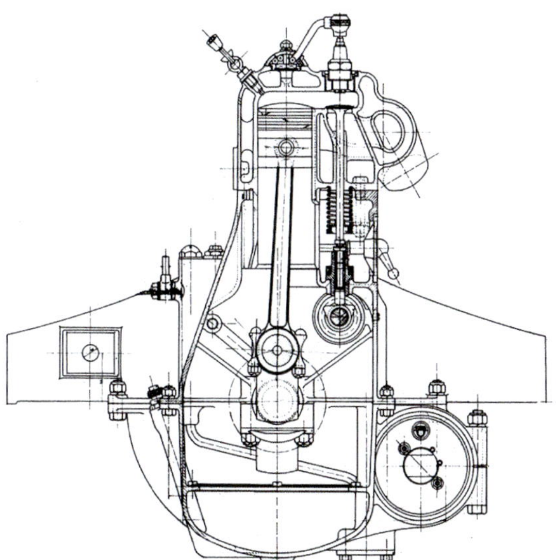

Fig. 5.4 Cross-section view of the engine FIAT 101 showing the side-valve arrangement

cover to allow easier periodic maintenance operations, including grinding the valves (Table 5.1). The FIAT 101 on display (Fig. 5.3) was used for research activities up to the 1940s, at the engine laboratory of the University of Palermo.

5.2 FIAT 114

1929–1932

The 114 engine was developed at the end of the 1920s for the FIAT 514 (Fig. 5.5), a compact car model, produced with different bodywork in large series from 1929 to 1932. The same engine was also used for the larger FIAT 515 model, built from 1931 until 1934. Unlike other FIAT engines, in the 114 model (Fig. 5.6) the four cylinders are grouped in a single cast iron block integrated with the crankcase while the oil sump is made of lightweight pressed sheet metal (Table 5.2). The combustion chamber features a common flat shape with side valves and removable head cover

Table 5.2 Technical specifications of the FIAT 114

Manufacturer	FIAT, Turin, Italy, from 1929 to 1932
Type	114
Description	Inline 4-cylinder engine, cylinder block made of cast iron with 3 main bearings for crankshaft (Fig. 5.7)
Stroke	102 mm
Bore	67 mm
Displacement	1438 cm^3
Compression ratio	5.85
Power	28 hp at 3400 rpm
Valvetrain	2 side valves per cylinder, controlled by a camshaft housed in the crankcase, driven by a Morse chain
Fuel feeding	Single barrel petrol carburettor Solex 26 FHD
Ignition	Single spark plug per cylinder supplied by an inductive discharge system with a high-voltage distributor
Cooling	Liquid-fed by a centrifugal pump
Lubrication	Forced by a gear pump
Application	FIAT 514 and FIAT 515 (various bodywork models)

Fig. 5.5 A FIAT 514 powered by the engine FIAT 114

allowing easy periodic maintenance operations, such as valve grinding (Fig. 5.7). Respect to many competitors, the ignition system is powered by distributor rather than magneto, a choice not very spread in the 1920s.

Fig. 5.6 The automotive engine FIAT 114, 1929–1932

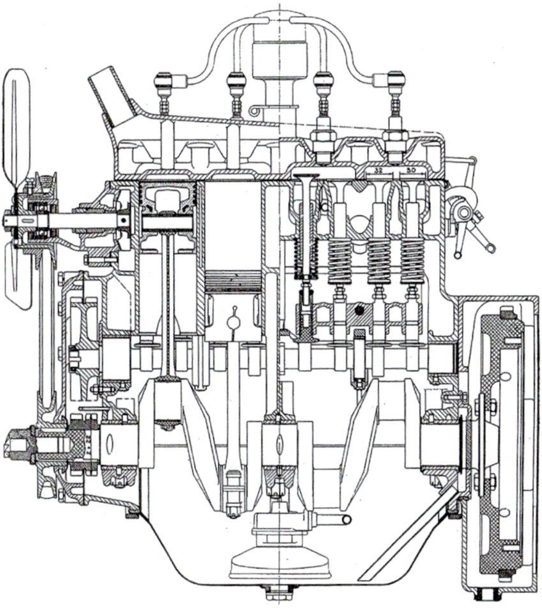

Fig. 5.7 Cross-section view of the automotive engine FIAT 114

5.3 Frera 250 VL

1926–1932

Founded in Tradate (Varese) by Corrado Frera in 1904, the Società Anonima Frera was one of the first Italian motorcycle companies and among the largest of the time for its production volume. On the eve of the First World War, its production included various models of motorcycles with single- and twin-cylinder engines; furthermore almost all Frera models could be transformed into military vehicles. After the war, Frera expanded and diversified its production, participating at the same time, directly and indirectly, in sporting competitions, with various successes. Between 1926 and 1932 Frera developed numerous models, including the Sport K3 with a 250 cc engine (Table 5.3), but in 1933, also due to the effects of the 1929 economic crisis, it was forced to close the factories. In the following years there were various attempts to resume production but due to modest results Frera disappeared towards the end of the 1950s.

Table 5.3 Technical specifications of the Frera 250 VL

Manufacturer	Società Anonima Frera, Tradate (Varese), Italy, from 1926 to 1932
Type	250 cc (Fig. 5.8)
Description	Single-cylinder engine, aluminium alloy crankcase, cylinder block and head made of cast iron
Stroke	88 mm
Bore	60 mm
Displacement	249 cm^3
Power	9 hp at 4600 rpm
Valvetrain	2 side valves, controlled by two gear-driven cams housed in the crankcase
Fuel feeding	Single barrel petrol carburettor
Ignition	Single spark plug supplied by a Magneti Marelli high-voltage magneto
Cooling	Air-fed
Lubrication	Open-loop oiling enforced by a manual piston pump with flow regulator
Application	Frera Sport 250 K3 (Fig. 5.9)

Fig. 5.8 The automotive engine Frera 250 VL, partially sectioned for educational purpose. Kind donation of Mr. Girolamo Di Giovanni

Fig. 5.9 The Frera Sport 250 K3 powered by the engine 250 cc VL

5.4 Isotta Fraschini 8 A

1924–1932

In 1919, after the reconversion of a large part of its wartime production, Isotta Fraschini introduced a new chassis conceived for various types of luxury cars to be bodywork fitted by the most renowned coachbuilders. Thanks to these valuable artisanal constructions, the new model, called type 8 from its 8-cylinders engine, achieved an international widespread success, becoming one of the automotive status symbols of the 1920s, in direct competition with the corresponding models produced by the most important manufacturers, such as Bugatti, Rolls-Royce and Duesemberg. In 1924 the 8 A model was introduced, with various developments to chassis and engine, including a larger displacement of 7.4 litres and a max. power of 115 hp (Table 5.4). This was shortly followed by the more powerful AS and ASS sports versions and the ultimate 160 hp model named 8 B. Due to 1929 economic crisis, the Isotta Fraschini's car production was interrupted in 1932, with an estimate total number of circa 1800 8-cylinder engine chassis built. Over time many of these have been scrapped, in some cases reusing the engines for other applications. The 8 A engine on display, serial no. 1474, was built between 1929 and 1930 for the chassis no. 1368. After WWII it was reused as emergency auxiliary power unit of the Trapani-Erice cableway in Sicily. Inaugurated in 1956, it remained in service until the beginning of the 1980s. In 2005 the old cableway was replaced by a modern system while the old driving station in Erice was abandoned with its equipment. Finally, in 2018 the Isotta Fraschini engine was recovered by the Museum, which carried out a complete restoration (Fig. 5.10).

Table 5.4 Technical specifications of the Isotta Fraschini 8 A

Manufacturer	Isotta Fraschini, Milano, from 1924 to 1932
Type	8 A
Description	Inline 8-cylinder engine, aluminium alloy crankcase with 9 crankshaft main bearings, head and cylinder block made of cast iron (Fig. 5.11)
Stroke	130 mm
Bore	98 mm
Displacement	7372 cm^3
Power	115 hp at 2500 rpm
Valvetrain	2 overhead valves per cylinder, operated by pushrods and rocker arms, driven by a single camshaft housed in the crankcase and driven by a Morse chain
Fuel feeding	Two single-barrel petrol carburettors Zenith HAK 36
Ignition	Single spark plug per cylinder supplied by a high-voltage Bosch distributor
Cooling	Liquid-fed by a centrifugal pump
Lubrication	Forced by a gear pump
Application	Isotta Fraschini 8A luxury car

Fig. 5.10 The Isotta Fraschini 8 A engine, at the beginning (left) and during its restoration at the Museum's workshop (right). Entrusted by the city of Erice

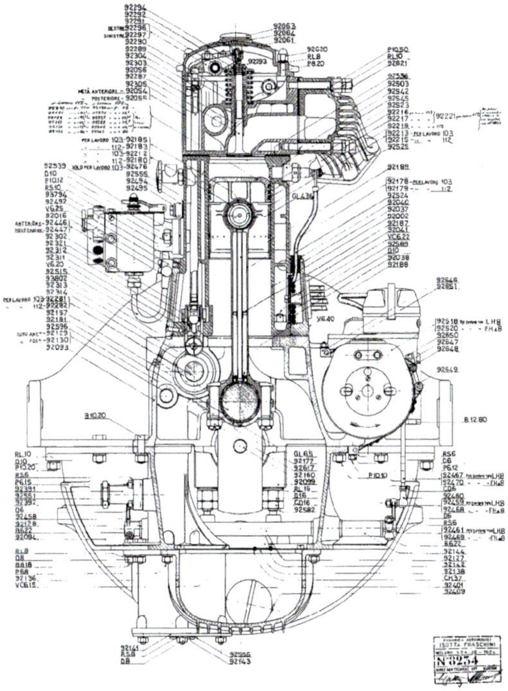

Fig. 5.11 Cross-section view of the automotive engine Isotta Fraschini 8 A

5.5 FIAT 350 C

1933

The 350 C is a Diesel engine produced by FIAT in the 1930s to equip its range of heavy trucks 632 N1, 633 N and other models derived from them, widely used for civil and military applications (Fig. 5.12). Given the intended use, the engine has been designed with a particularly robust structure, allowing easy repair and maintenance operations. Its overall layout (Table 5.5) was used by FIAT as a development basis for the subsequent engine models with higher displacement and six-in-line cylinders. The engine (Fig. 5.13) is equipped with an in-cylinder direct injection system, with a combustion chamber of modern design, obtained by a particular shape of the piston crown (Fig. 5.14).

Fig. 5.12 A heavy truck FIAT 632, 1933

Table 5.5 Technical specifications of the FIAT 350 C

Manufacturer	FIAT, Turin, 1933
Type	350 C
Description	Inline 4-cylinder 4-stroke Diesel engine, cylinder block and head made of cast iron, aluminium alloy oil sump, crankshaft with 5 main bearings (Fig. 5.14)
Stroke	152 mm
Bore	108 mm
Displacement	5570 cm^3
Compression ratio	15.0
Power	60 hp at 1800 rpm
Valvetrain	2 overhead valves pe cylinder, controlled by pushrods and rocker arms by a single camshaft housed in the crankcase and driven by gears and triple chain
Fuel feeding	in-cylinder direct injection supplied by a mechanical piston pump with centrifugal speed regulator, max. feeding pressure 150 bar
Cooling	Liquid-fed by a centrifugal pump
Lubrication	Forced by a gear pump
Application	Heavy truck FIAT 632 N1 (1933) and 633 N (1934)

Fig. 5.13 The heavy truck engine FIAT 350 C, 1933. Entrusted by Stellantis

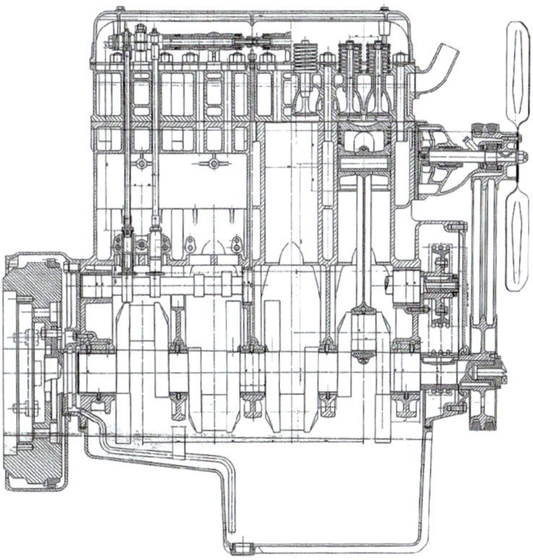

Fig. 5.14 Cross-section view of the 4-stroke Diesel engine FIAT 350 C

5.6 Lancia-Junkers type 89—*Lancia Ro*

1932–1938

At the beginning of the 1930s, Lancia purchased the manufacturing license of a particular engine developed in Germany by Junkers: a two-stroke Diesel, with two opposing pistons in each cylinder. The intake and exhaust ports are located at the upper and lower ends, near the opposite dead centres of the piston stroke, making a high-efficiency uniflow-scavenging-type combustion chamber. Moreover, upper pistons are made with have two different diameters since the top side ones, having larger bore, work as air pump. Each upper piston is connected to the crankshaft by two long connecting rods, which in turn co-support axial thrusts originates from in-cylinder combustion and part of the inertia forces. As a result, this arrangement allows a compact and lightweight design. Named Type 89, the engine was produced from 1932 to 1938 for the Diesel fuel version of the heavy truck Lancia Ro (Fig. 5.15 and Table 5.6).

Table 5.6 Technical specifications of the Lancia-Junkers type 89

Manufacturer	Lancia (under license of Junkers), Turin, Italy, from 1932 to 1938
Type	89 (Fig. 5.16)
Description	Inline 2-cylinder two-stroke Diesel engine, with two counter-swinging pistons in each cylinder, crankcase and engine block made of aluminium alloy, with steel-made cylinder liners (Fig. 5.17)
Stroke	100 mm (upper pistons), 150 mm (lower piston)
Bore	90 mm
Displacement	3180 cm^3
Power	64 hp at 1500 rpm
Specific power	20.1 hp/litre
Air intake	Uniflow scavenging forced by piston pump
Fuel feeding	in-cylinder injection supplied by a 2-cylinder reciprocating Diesel fuel pump, with max. feeding pressure of 500 bar
Cooling	Liquid-fed by a centrifugal pump
Lubrication	Forced by a gear pump
Application	Military and civil heavy truck Lancia Ro (Fig. 5.15)

Fig. 5.15 A heavy truck Lancia Ro powered by Lancia-Junkers type 89 engine, 1932–1938

Fig. 5.16 The heavy truck two-stroke Diesel engine Lancia-Junkers type 89, 1932–1938

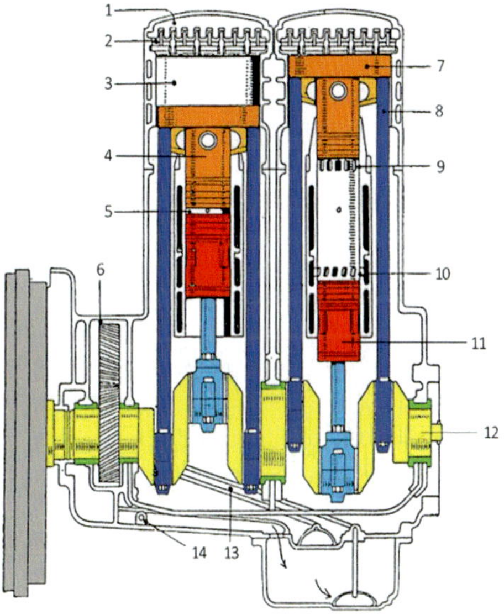

Fig. 5.17 Schematic cross-section view of Lancia-Junkers type 89 engine. Legend: (1) air intake box; (2) air inlet automatic valve; (3) air pump cylinder; (4) upper piston; (5) combustion chamber; (6) fuel injection gear drive; (7) air pump piston; (8) upper connecting rod; (9) in-cylinder air inlet port; (10) exhaust port; (11) lower piston; (12) crankshaft; (13) oil feeding duct; (14) oil scavenge

5.7 Benelli 500 VL

1940

The 500 VL engine (VL stands for side valves) was introduced in 1940 for the Benelli 500 VLC motorcycle (Fig. 5.18). Its production, initially conceived for the sport, was then quickly and primarily focused for military application (Table 5.7). To this purpose, a side-valve model (Fig. 5.19), named 500 VLM, was specifically developed since this arrangement was considered more reliable than the sportier overhead valve head.

Table 5.7 Technical specifications of the Benelli 500 VL

Manufacturer	Benelli, Pesaro, Italy, 1940
Type	500 VL
Description	Single-cylinder engine, cast iron cylinder block, head and crankcase made of aluminium alloy
Stroke	87 mm
Bore	85 mm
Displacement	493 cm^3
Compression ratio	4.1
Power	12 hp at 4750 rpm
Valvetrain	2 side valves, controlled by two gear-driven cams housed in the crankcase
Fuel feeding	Single-barrel petrol carburettor Dell'Orto MC 26 F
Ignition	Single spark plug supplied by a high-voltage magneto
Cooling	Air-fed
Lubrication	Forced by a gear pump (Fig. 5.20)
Application	Benelli 500 VLC (Civilian) and 500 VLM (Military applications)

Fig. 5.18 The motorcycle Benelli 500 VL, 1940

Fig. 5.19 The motorcycle engine Benelli 500 VL, 1940

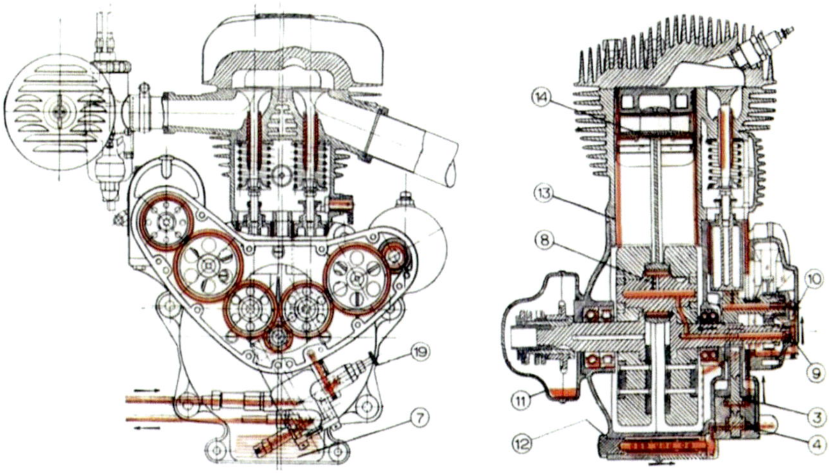

Fig. 5.20 Cross-section views of the engine Benelli 500 VL, highlighting the lubrication system, cams and accessories gear train and the side-valve combustion chamber

5.8 Innocenti Lambretta 125 C

1950–1951

After the Second World War, Innocenti, Piaggio and other Italian companies reconverted their wartime production starting the development of light motorcycles and scooters, which gained a significant role into the motor vehicle widespread in Italy, as well as to the economic rebirth of the country. In 1947 Innocenti presented the first version of its famous and innovative scooter, named Lambretta, which was mass produced in various series until 1972 (Fig. 5.21). Thanks to its versatility and affordable cost, it achieved a rapid commercial success and had a wide diffusion in many countries, becoming an icon of popular culture like its main contemporary competitor, the Piaggio Vespa. Respect to the first models, the 1950 125 C (Fig. 5.22) features some improvements, including the introduction of the rear suspension system. In its

Fig. 5.21 The Innocenti assembly line of the Lambretta scooters in the 1950s

Table 5.8 Technical specifications of the Innocenti Lambretta 125 C

Manufacturer	Innocenti, Milan, Italy, from 1950 to 1951
Type	125 C
Description	Single-cylinder 2-stroke engine, cast-iron cylinder, piston, head and crankcase made of aluminium alloy
Stroke	58 mm
Bore	52 mm
Displacement	123 cm^3
Compression ratio	6.0
Power	4.3 hp at 4500 rpm
Air intake	Crossflow scavenging forced by crankcase compression
Fuel feeding	Single-barrel petrol carburettor Dell'Orto MA16
Ignition	Single spark plug supplied by a high-voltage magneto
Cooling	Air-fed
Lubrication	Supplied by a premixed fuel-oil mixture (5%)
Clutch	Oil-wet multiple-plate type
Transmission	3-speed gearbox
Application	Lambretta 125 C (Fig. 5.23) and 125 LC

particular arrangement the engine frame is a single assembly with gearbox and, final transmission and rear suspension (Table 5.8).

Fig. 5.22 The motorcycle engine Innocenti Lambretta 125 C, 1950–1951. The engine on display, partially sectioned for educational purposes, includes all powertrain components: the multi-plate clutch, the gearbox, the rear wheel suspension with shock absorber and drum brake

Fig. 5.23 An Innocenti Lambretta 125 C, 1950–1951

5.9 Moto Guzzi Galletto 160 c.c.

1950–1952

In the late 1940s Moto Guzzi developed a new light vehicle, named Galletto, with intermediate characteristics and performance between a motorcycle and a scooter. The prototype was presented at the Geneva Motor Show in 1950 and the production of the first series, equipped with a 160 cm3 four-stroke engine (Fig. 5.24) with an integrated 3-speed gearbox, was started in the same year. In 1952 the second series was introduced with a larger 175 cm3 engine (192 cm3 from 1954), 4-speed gearbox and various technical improvements. The third and last series was introduced in 1961 and it was kept in production up to 1966, when the ownership of Moto Guzzi passed from the Parodi family to the SEIMM group (Table 5.9).

Table 5.9 Technical specifications of the Moto Guzzi Galletto 160 c.c. powertrain

Manufacturer	Moto Guzzi, Mandello del Lario (Lecco), from 1950 to 1952
Type	Galletto 160 c.c.
Description	Single-cylinder 4-stroke engine, cast-iron cylinder block, head and crankcase made of aluminium alloy (Fig. 5.24)
Stroke	53.0 mm
Bore	62.0 mm
Displacement	159.5 cm^3
Compression ratio	5.6
Power	6 hp at 5200 rpm
Specific power	37.6 hp/litre
Valvetrain	2 overhead valves per cylinder, operated by pushrods and rocker arms driven by a camshaft housed in the crankcase
Fuel feeding	Single-barrel petrol carburettor Dell'Orto
Ignition	Magneto flywheel Filso or Marelli
Cooling	Air-fed
Lubrication	Forced by a gear pump
Clutch	Oil-wet multiple-plate type
Transmission	3-ratio sequential manual gearbox integrated into the crankcase
Application	Light motocycle Moto Guzzi Galletto 160 c.c. (Fig. 5.25)

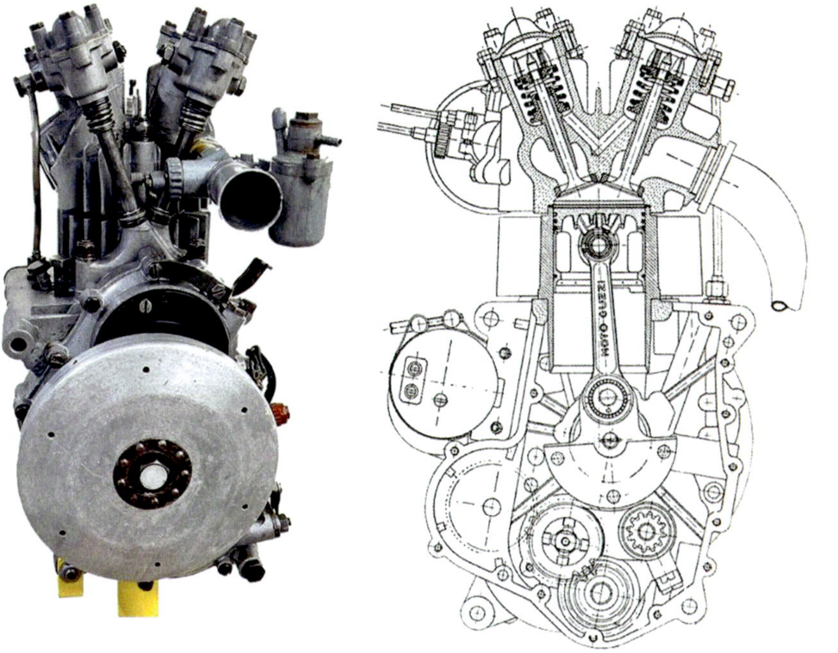

Fig. 5.24 The engine Moto Guzzi Galletto 160 c.c. (left) with its cross-section view (right). Kind donation of Mr. Gilberto Bonafede

Fig. 5.25 The motorcycle engine Moto Guzzi Galletto 160 c.c., 1950–1952

5.10 FIAT 104–*8V*

1952–1954

The type 104 (Fig. 5.26) is a high-performance engine developed by FIAT for its sport car named 8V. Due to its fine design (Table 5.10) it is a rare and particular model among the FIAT production at the beginning of the 1950s. The 8V were manufactured at the FIAT experimental workshop, for a total production of only 114 units between 1952 and 1954. Many chassis-engine assemblies were sold to independent coach-builder which produced special versions, such as the FIAT 8V Zagato (Fig. 5.27) and the Siata 208. A good number of 8V, especially those manufactured by Zagato, were employed successfully in numerous sport competitions. Among these are worthy of note: the victory of the Targa Florio in 1955 (1st 8V Zagato, 2nd FIAT 8V), the first position achieved at the AVUS circuit in the Grand Prix of Berlin in 1955 (8V Zagato) and the victory of five consecutive Italian Gran Turismo championships from 1954 to 1958 (8V Zagato).

Table 5.10 Technical specifications of the FIAT 104–8V

Manufacturer	FIAT, Turin, Italy, from 1952 to 1954
Type	104.000
Description	V-70° 8-cylinder engine, cylinder block and head made of aluminium alloy, flat-plane crankshaft with three main bearings
Stroke	61.3 mm
Bore	72 mm
Displacement	1996 cm^3
Compression ratio	8.5
Power	105 hp at 5600 rpm
Torque	146 Nm at 3600 rpm
Specific power	52.6 hp/litre
Valvetrain	2 overhead valves per cylinder, operated by pushrods and rocker arms driven by single camshaft housed between the cylinder banks
Fuel feeding	Two twin-barrel petrol carburettors Weber 36 DCZ 3
Ignition	Single spark plug per cylinder supplied by an inductive discharge system with a single high-voltage distributor
Cooling	Liquid-fed by a centrifugal pump
Lubrication	Forced by a gear pump, with oil cooler
Application	Sport car FIAT 8V and Siata 208

Fig. 5.26 The automotive engine FIAT 104–*8V*, 1952–1954

Fig. 5.27 FIAT 8V Zagato during a sport car competition in the 1950s

5.11 Breda 2 H 88

1950s

The Breda engine 2 H 88 (Fig. 5.28) derives from a project originally designed by the French engineer Roland Laraque in the 1940s, then built by the Manufacture d'Armes de Paris (M.A.P.) from 1947 to 1950 for the DR3 agricultural tractor. In Italy the engine was produced under license by Breda Motori up to 1959, for the Ansaldo Fossati FB4r tractor. From a technical point of view, the engine works according to the two-stroke Diesel cycle, and it is endowed with a Roots supercharger (Table 5.11). The engine features a particular compact arrangement made with only two horizontal cylinders. In each of them two opposing pistons work, connected to the crankshaft by means of two connecting rods and a central rocker arm (Fig. 5.29). The intake and exhaust ports are machined at the opposite dead points of each cylinder, making a high-efficiency uniflow-scavenging-type combustion chamber. The engine on display is partially sectioned for educational purposes.

Table 5.11 Technical specifications of the Breda 2 H 88

Manufacturer	Breda Motori S.p.A. (license of M.A.P.), Milan, 1950s
Type	2 H 88
Description	Inline 2-cylinder two-stroke Diesel engine, with two counter-swinging pistons per cylinder; double connecting rod and rocker arm for each piston; cast-iron cylinder block, aluminium alloy oil sump (Fig. 5.29)
Stroke	102 mm
Bore	88 mm
Displacement	2482 cm^3
Compression ratio	16
Power	50.3 hp at 1800 rpm
Specific fuel consumption	170 g/hp/h
Specific power	20.3 hp/litre
Air intake	Uniflow scavenging forced by a Roots supercharger, max. 1.4 bar boost pressure (absolute) at 1500 rpm
Fuel feeding	in-cylinder Diesel fuel injection supplied by a 2-cylinder reciprocating pump, with centrifugal speed regulator and max. feeding pressure of 110 bar
Cooling	Liquid-fed, with centrifugal pump
Lubrication	Forced by gear pump
Mass	330 kg (dry)
Mass-to-power ratio	6.56 kg/hp
Application	Agricultural tractors Ansaldo Fossati FB4r

Fig. 5.28 The two-stroke Diesel engine Breda 2 H 88, 1950s

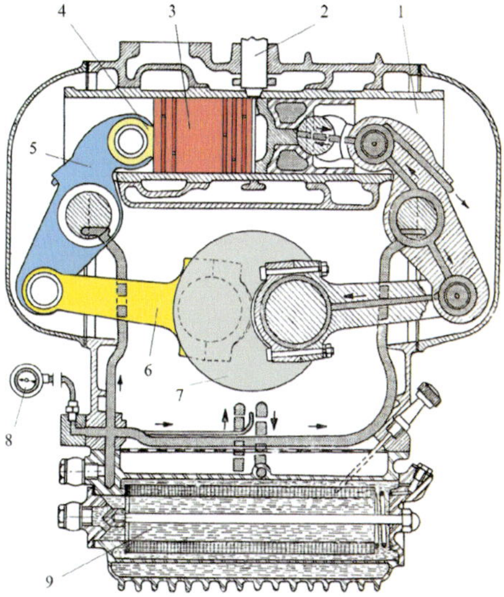

Fig. 5.29 Cross-section view of Breda 2 H 88: (1) cylinder; (2) fuel injector; (3) piston; (4) upper connecting rod; (5) rocker arm; (6) lower connecting rod; (7) crankshaft; (8) oil sump with filter

5.12 FIAT 305

1953–1968

The Diesel engine type 305 (Table 5.12) was developed by FIAT, in the early '50s, starting from the petrol engine type 105. Initially it was used for the light truck FIAT 615 N and, in 1953, for the FIAT 1400 Diesel (Fig. 5.30), making it the first Diesel engine mass-produced in Italy for passenger cars. From the same year the engine was also used for the light off-road vehicle FIAT Campagnola Diesel. The 305 engine was built for over 15 years in various versions, that on display has been partially sectioned for educational purposes (Fig. 5.31).

Fig. 5.30 The FIAT 1400 sedan, 1950s

Table 5.12 Technical specifications of the FIAT 305 powertrain

Manufacturer	FIAT, Turin, Italy, from 1955 to 1960
Type	305B.007
Description	Inline 4-cylinder engine, head and cylinder block made of cast iron, crankshaft with 3 main bearings
Stroke	90 mm
Bore	82 mm
Displacement	1901 cm^3
Compression ratio	20.0
Power	43 hp at 3200 rpm
Torque	96 Nm at 2000 rpm
Valvetrain	2 overhead valves per cylinder, controlled by pushrods and rocker arms operated by a single camshaft housed in the crankcase (Fig. 5.32)
Fuel feeding	Indirect injection system with Ricardo-Comet type prechamber, supplied by a 4-piston Diesel fuel pump
Cooling	liquid-fed by a centrifugal pump
Lubrication	Forced by a gear pump
Clutch	Single-disc type
Transmission	4 speed and reverse manual gearbox
Application	FIAT 615 N (truck), FIAT 1400 Diesel (sedan car), FIAT Campagnola Diesel (off-road vehicle)

Fig. 5.31 The automotive Diesel engine FIAT 305, 1953–1968

Fig. 5.32 Cross-section view of the Diesel engine FIAT 305

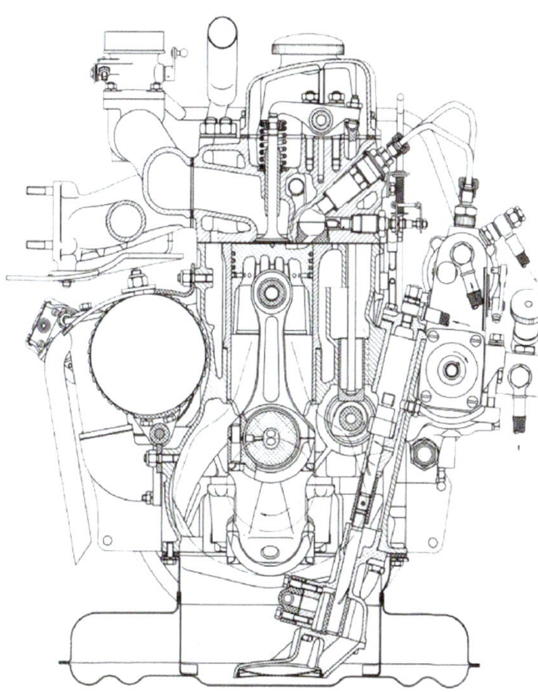

5.13 FIAT 103–*1100 TV*

1956–1957

In the second half of the 1930s, FIAT introduced with a new four-cylinder 1100 cm^3 engine featuring overhead valves and an innovative combustion chamber, designed by the Dante Giacosa, who was one of the most creative FIAT technical designers. Compared to the previous widespread side-valve engines, the compact shape with the spark plug in optimal position, allowed a faster and more efficient combustion. In short time, this arrangement was adopted by many other manufacturers for engines of same category. After the war the engine underwent several technical upgrades for better performance and reliability. The 103 model was developed for the 1950s FIAT 1100/103 cars (Fig. 5.33). The engine on display (Fig. 5.34) belongs to the

Fig. 5.33 The FIAT 1100 sedan, 1956–1957

Table 5.13 Technical specifications of the FIAT 103–*1100 TV*

Manufacturer	FIAT, Turin, Italy, from 1956 to 1957
Type	103 with high compression head "TV"
Description	Inline 4-cylinder engine, aluminium alloy head, cast-iron cylinder block, crankshaft with 3 main bearings
Stroke	75 mm
Bore	68 mm
Displacement	1089 cm^3
Compression ratio	8.0
Power	53 hp at 5400 rpm
Torque	71 Nm at 3600 rpm
Valvetrain	2 overhead valves per cylinder, operated by pushrods and rocker arms driven by single camshaft (Fig. 5.35)
Fuel feeding	Twin-barrel petrol Weber carburettor DCD 28/36
Ignition	Single spark plug per cylinder supplied by an inductive discharge system with Magneti Marelli distributor
Cooling	Liquid-fed by a centrifugal pump
Lubrication	Forced by a gear pump
Application	FIAT 1100/103 (sedan, spider et al.)

most powerful version, equipped with a "TV" head with higher compression ratio and twin-barrel carburettor (Table 5.13).

Fig. 5.34 The automotive engine FIAT 103–*1100 TV*, 1956–1957

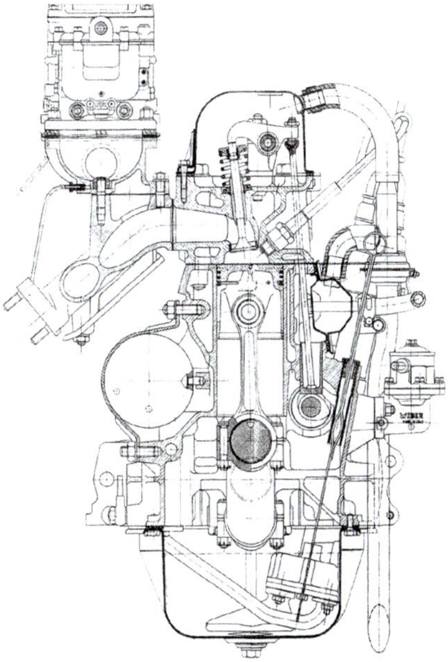

Fig. 5.35 Cross-section views of the FIAT engine type 103

5.14 Lancia 813–*Flaminia 2500*

1956–1961

In 1950 Lancia was the first automotive manufacturer to put into series production a V-60° six-cylinder engine, which powered the Aurelia sedan and coupé models. That original cylinder arrangement, nowadays widely used by many manufacturers, was developed during the 1940s by Franco De Virgilio, head of the Special Studies and Patent Office at Lancia. The engine design features an extensive use of aluminium alloys and a fine transaxle transmission (Fig. 5.36), used for a better on-board mass balance, which entails higher vehicle dynamic performances. Over the years the V-6 engine underwent continuous technical upgrades, such as lager displacements, which resulted in higher power outputs for larger or sportier cars. The engine model 813 was introduced in 1957 for the new Lancia Flaminia sedan which was produced in different versions up to 1970 with many engine upgrades (Table 5.14).

Table 5.14 Technical specifications of the engine Lancia 813 powertrain

Manufacturer	Lancia, Turin, Italy, from 1957 to 1961
Type	813.00
Description	V-60° 6-cylinder engine, cylinder block, heads and oil pan made of aluminium alloy, cast-iron cylinder liners
Stroke	81.5 mm
Bore	80 mm
Displacement	2458 cm^3
Compression ratio	7.8
Power	102 hp at 4600 rpm
Torque	183 Nm at 3500 rpm
Specific power	41.5 hp/litre
Valvetrain	2 overhead valves per cylinder, operated by pushrods and rocker arms driven by a single camshaft housed between the cylinder banks (Fig. 5.37)
Fuel feeding	Twin-barrel petrol carburettor Solex C35 PAAI
Ignition	Single spark plug per cylinder supplied by an inductive discharge system with a high-voltage distributor, firing order: 1-4-3-6-5-2
Cooling	Liquid-fed by a centrifugal pump
Lubrication	Forced by a gear pump, with an oil cooler
Transmission	Transaxle-type with clutch, gearbox and differential gear train housed into an integrated aluminium alloy assembly at the rear axle;
Clutch	Single-disc type
Gearbox	4 speed and reverse
Application	Lancia Flaminia 2500 sedan

Fig. 5.36 The Lancia Flaminia 2500 powertrain with the engine type 813, 1956–1961

Fig. 5.37 Cross section view a the Lancia V-60° six-cylinder engine

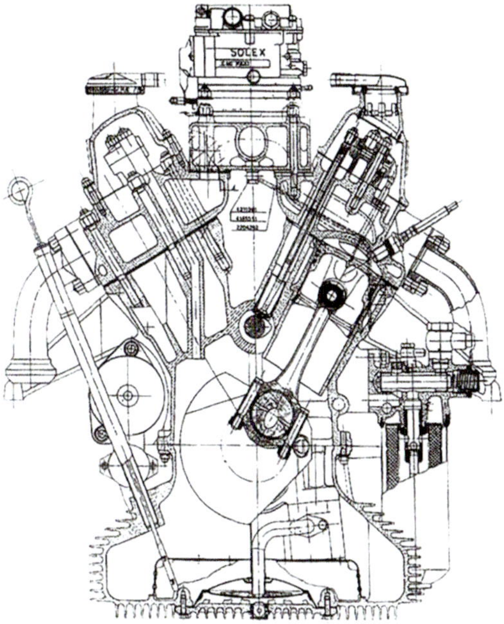

5.15 FIAT-OM CO2D

1957–1972

The Officine Meccaniche (OM) of Brescia was a famous mechanical factory, whose origins date back to the first half of the 19th century. At the beginning of the 20th century, production was expanded to the automotive sector, with the creation of various models and participation in numerous sporting competitions. After 1933, as a consequence of the company's acquisition by FIAT, OM concentrated its activities for industrial and heavy vehicles, until it became part of FIAT Veicoli Industriali in 1967, losing its design autonomy. The CO2D Diesel engine (Table 5.15), produced by FIAT-OM from 1957 to 1972 for the OM truck Tigrotto (Fig. 5.38), dates back to that period. The engine layout, like FIAT models, features a robust cast-iron structure with some components in aluminium alloy (Figs. 5.39 and 5.40) and an in-cylinder direct injection system, with part of the combustion chamber created in the crown of the pistons.

Fig. 5.38 The light truck OM Tigrotto powered by the Diesel engine FIAT-OM CO2D

Table 5.15 Technical specifications of the FIAT-OM CO2D

Manufacturer	Officine Meccaniche (OM), Brescia, Italy, 1957-1972
Type	CO2D—10
Description	Inline 4-cylinder engine, head and cylinder block made of cast iron, crankshaft with 5 main bearings
Stroke	120 mm
Bore	108 mm
Displacement	4397 cm^3
Compression ratio	16
Power	85 hp at 2400 rpm
Torque	258 Nm at 1200 rpm
Valvetrain	2 overhead valves per cylinder, operated by pushrods and rocker arms driven by a camshaft in the crankcase
Fuel feeding	in-cylinder direct injection system, single injector per cylinder supplied by a 4-piston Diesel fuel pump
Cooling	Liquid-fed by a centrifugal pump
Lubrication	Forced by a gear pump
Application	Light truck OM Tigrotto (Fig. 5.38)

Fig. 5.39 The light truck Diesel engine FIAT-OM CO2D, 1957–1972

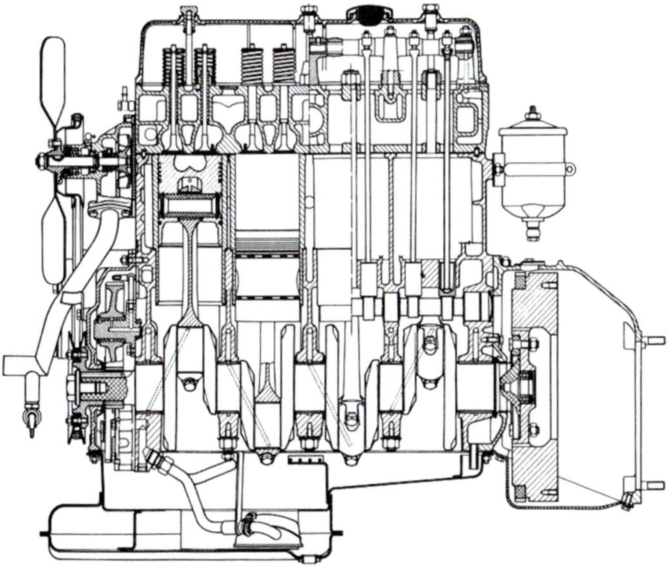

Fig. 5.40 Cross-section view of the light truck Diesel engine FIAT-OM CO2D

5.16 FIAT 213

1959–1964

The 213 is a straight-six engine conceived and produced by FIAT at the end of the 1950s for Diesel fuel (N) versions of its range of medium-light trucks, such as FIAT C 40 N (Fig. 5.41) and FIAT C 50 N. To this purpose, the engine features a robust and reliable design (Table 5.16) with great attention for easy maintenance operations. The same general arrangement was widely used as a common base for various smaller four-cylinder models. To reduce the overall mass, some components, including the large oil pan and the clutch bell, are made of aluminium alloy. Similarly to many contemporary engines, the 213 model (Fig. 5.42) is equipped with a common in-cylinder direct injection system, with part of the combustion chamber created in the crown of the pistons (Fig. 5.43).

Table 5.16 Technical specifications of the FIAT 213

Manufacturer	FIAT, Turin, Italy, from 1959 to 1964
Type	213.000
Description	Inline 6-cylinder engine, head and cylinder block made of cast iron, aluminium alloy oil sump, crankshaft with 7 main bearings (Fig. 5.43)
Stroke	110 mm
Bore	95 mm
Displacement	4678 cm^3
Compression ratio	16.9
Power	100 hp at 3000 rpm
Torque	257 Nm at 1700 rpm
Valvetrain	2 overhead valves per cylinder, operated by pushrods and rocker arms driven by a camshaft in the crankcase
Fuel feeding	in-cylinder direct injection system, single injector per cylinder supplied by a 6-piston Diesel fuel pump
Cooling	Liquid-fed by a centrifugal pump
Lubrication	Forced by a gear pump
Application	FIAT C 40 N (Fig. 5.41), FIAT C 50 N et al.

Fig. 5.41 The light truck FIAT C 40 N powered by the Diesel engine FIAT 213

Fig. 5.42 The light truck Diesel engine FIAT 213, from 1959 to 1964

Fig. 5.43 Cross-section view of the light truck Diesel engine FIAT 213

5.17 Alfa Romeo 00100–*1300 Twin-Camshaft*

1960–1961

After the Second World War, Alfa Romeo gradually renewed its automotive production range with the development of new models. To this purpose, starting from 1950 a new four-cylinder engine was developed, for various applications. It was characterized by the extensive use of aluminium alloys, an open-deck cylinder block with inserted cast-iron cylinders and hemispherical combustion chambers with V-90° overhead valves. Due to its technical features (Table 5.17) and good performance, it achieved a widespread commercial and racing success, becoming famous as *bialbero* (i.e. twin-camshaft), and it was the basis of a long series of engines produced with different displacements and continuous upgrades for over 30 years. The 00100 model was introduced in 1960 for the Giulietta sedan (Figs. 5.44 and 5.45) and it was then used, with various modifications, for other models, including commercial vehicles (Table 5.17).

Fig. 5.44 The Alfa Romeo Giulietta sedan, 1960–1961

Table 5.17 Technical specifications of the Alfa Romeo 00100–*1300 twin-camshaft*

Manufacturer	Alfa Romeo, Milan, Italy, from 1960 to 1961
Type	AR 00100
Description	Inline 4-cylinder engine, head, oil sump and cylinder block made of aluminium alloy, steel cylinder liners, crankshaft with 5 main bearings (Fig. 5.46)
Stroke	75 mm
Bore	74 mm
Displacement	1290 cm^3
Compression ratio	8.5
Power	62 hp at 6500 rpm
Valvetrain	2 overhead valves per cylinder, operated by two overhead camshafts driven by roller-chain and gears
Fuel feeding	Single-barrel petrol carburettor Solex C 32 BIC
Ignition	Single spark plug per cylinder supplied by an inductive discharge system with a high-voltage distributor
Cooling	Liquid-fed by a centrifugal pump
Lubrication	Forced by a gear pump
Application	Alfa Romeo Giulietta sedan (1960-1961 model), Alfa Romeo Giulietta Promiscua (station wagon, 1960)

Fig. 5.45 The automotive engine Alfa Romeo 00100–1300 *twin-camshaft*, 1960–1961

Fig. 5.46 Cross-section view of the automotive engine Alfa Romeo 00100–*1300 twin-camshaft*

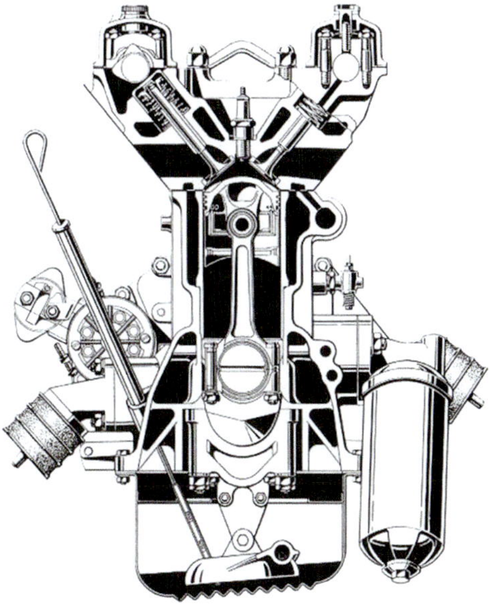

5.18 FIAT 116

1961–1966

The 116 is part of a large family of engines with different displacements and with 4 or 6 in-line cylinders, designed by FIAT in the 1960s for various medium-high category cars. From a technical point of view, these engines share the same overall arrangement with a cast-iron crankcase and an aluminium alloy head (Table 5.18). The latter features a particular shape of the combustion chamber, named *polyspherical*, created by a particular inclination of exhaust and intake valves. The model 116 is the entry level version used for the FIAT 1300 sedan with a 4-speed gearbox (Figs. 5.47 and 5.48).

Table 5.18 Technical specifications of the FIAT 116 powertrain

Manufacturer	FIAT, Turin, Italy, from 1961 to 1966
Type	116.000
Description	Inline 4-cylinder engine, aluminium alloy head, cast-iron cylinder block, crankshaft with 3 main bearings
Stroke	79.5 mm
Bore	72 mm
Displacement	1295 cm^3
Compression ratio	8.8
Power	65 hp at 5200 rpm (Fig. 5.49)
Torque	93 Nm at 3400 rpm
Valvetrain	2 overhead valves per cylinder, operated by pushrods and rocker arms driven by a camshaft housed in the crankcase (Fig. 5.49)
Fuel feeding	Single twin-barrel petrol Solex carburettor C28-32
Ignition	Single spark plug per cylinder supplied by an inductive discharge system with high-voltage distributor
Cooling	Liquid-fed by a centrifugal pump
Lubrication	Forced by a gear pump
Clutch	Single-disc
Transmission	4-speed and reverse manual gearbox
Application	FIAT 1300 sedan (Fig. 5.47) and station wagon

Fig. 5.47 The FIAT 1300 sedan, 1961–1966

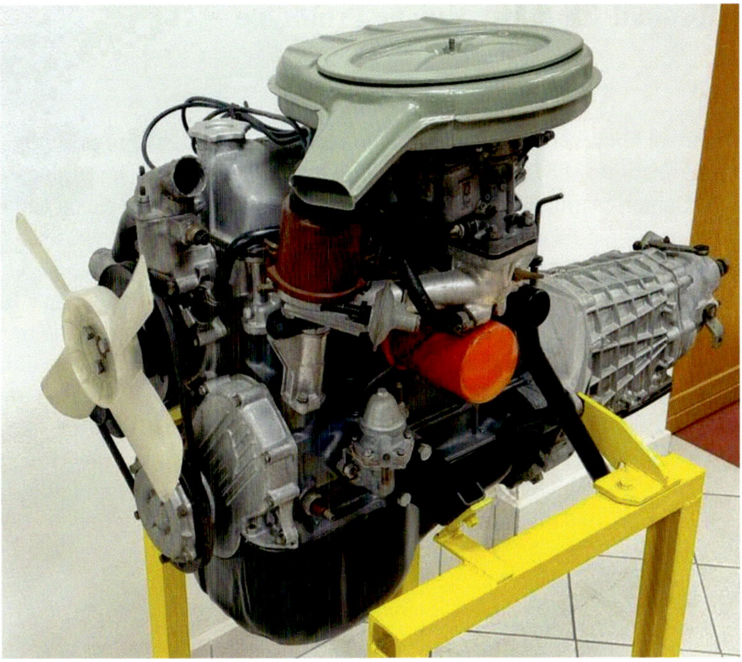

Fig. 5.48 The automotive engine FIAT 116 with the 4-speed gearbox, 1961–1966

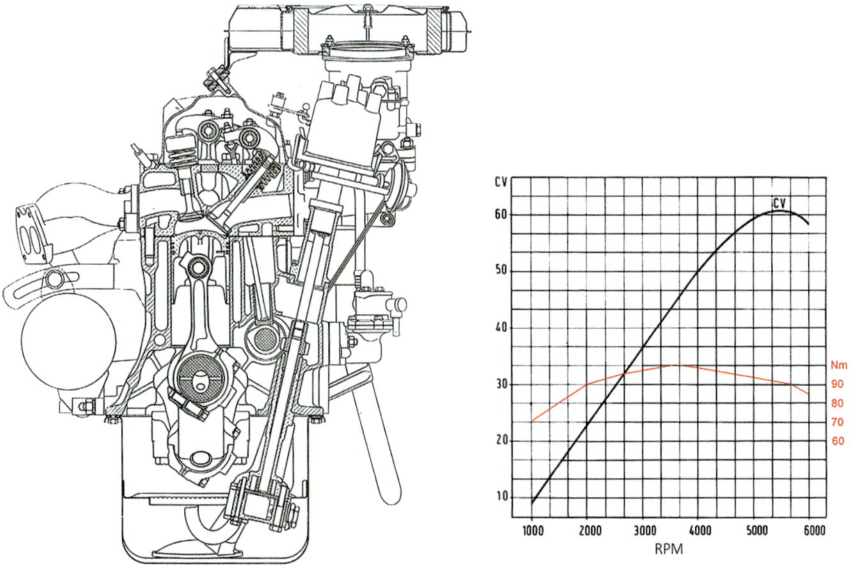

Fig. 5.49 Cross-section view of the engine FIAT 116 (left) with power and torque curves (right)

5.19 Toroidal-Chamber Engine Prototype

1961–1962

This two-stroke Diesel engine prototype was studied and designed in the early 1960s at the former Institute of Machinery of the University of Palermo. The engine is made up of two counter-oscillating parts, each of which has three 120° spaced truncated-toroidal piston (Fig. 5.50). The two parts together form 6 toroidal combustion chambers performing an ordinary two-stroke Diesel cycle (Figs. 5.51 and 5.52). The two counter-oscillating bodies are reciprocally connected by a bevel-gear mechanism driving the power shaft. Each counter-oscillating bodies is connected to the frame by means of an external roller ring "freewheel" mechanism, which ensures the unidirectional driving to the respective bevel gears (Table 5.19). Despite a certain reduction in the overall dimensions of the engine, this arrangement entails severe impulsive stresses greater than an ordinary piston-connecting rod mechanism. This system is one of numerous "unconventional" arrangements that have been studied in the development of internal combustion engines.

Table 5.19 Technical specifications of the toroidal-chamber engine prototype

Description	Two-stroke Diesel engine prototype, with counter-swing-pistons in 6 toroidal combustion chambers (Figs. 5.50 and 5.51)
Displacement	2850 cm^3
Fuel feeding	In-cylinder injection supplied by two single-cylinder reciprocating fuel pumps (one for three injectors)
Cooling	Air-fed
Lubrication	Forced by a mechanical pump
Application	Originally conceived for trucks and industrial vehicles

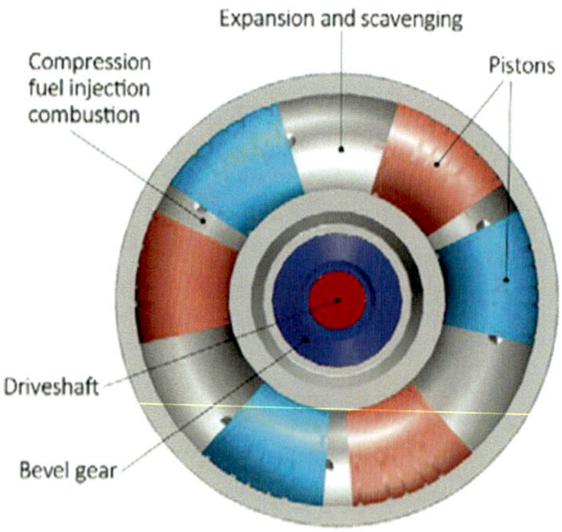

Fig. 5.50 Schematic cross-section view of the prototype of a two-stroke Diesel toroidal engine

Fig. 5.51 The prototype of a two-stroke Diesel toroidal engine, 1961–1962

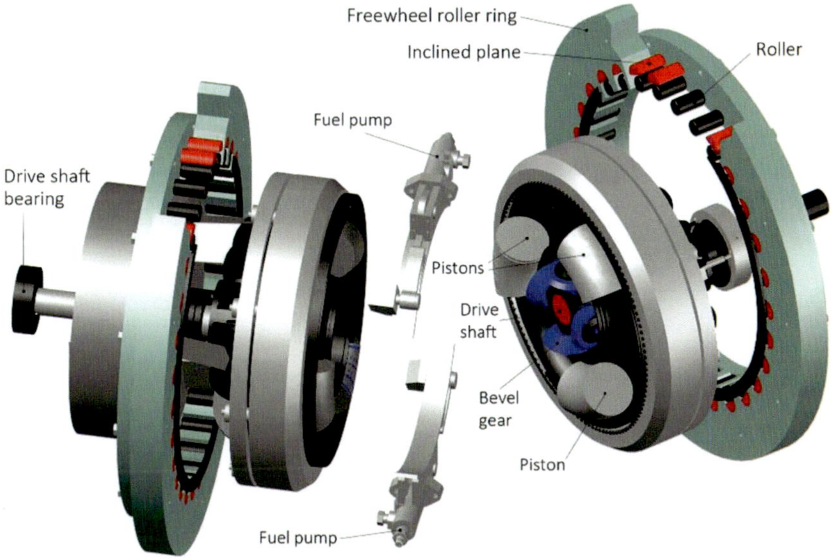

Fig. 5.52 Schematic view of the prototype of the two-stroke Diesel toroidal engine

5.20 Lancia 818.100–*Fulvia 2C*

1964–1969

In 1963 Lancia presented the new Fulvia sedan to replace the third and final series of the Appia. The new front-wheel drive car, characterized by the all-forward layout of its powertrain, required the design of a new and compact four-cylinder engine, named type 818.000. To this purpose, due to the limited space available, a narrow V-shaped cylinder arrangement of a bit more than 12 degrees was specifically designed. This allowed the use of a single head for all cylinders, reducing the size of the engine, which was also inclined to the right side of the vehicle to lower the front bonnet overall height. Thanks to various design upgrades and refinements, including the increase in engine displacement for greater power output, the engine was produced in many series for the Fulvia sedan and for the Fulvia coupé up to 1976. Both Lancia and numerous private developers created various racing models, which were successfully used in many competitions. The engine on display (Fig. 5.53), built in 1966, belongs to the second construction series, named 818.100, introduced in 1964 for the Fulvia 2C sedan. It differs from the first series in the use of two horizontal twin-barrel carburettors instead of a single carburettor, which was replaced for better performance (Table 5.20).

Table 5.20 Technical specifications of the Lancia 818.100–*Fulvia 2C*

Manufacturer	Lancia, Turin, Italy, from 1964 to 1969
Type	818.100
Description	V-12° 4-cylinder engine, single aluminium alloy head, cast-iron cylinder block, crankshaft with 3 main bearings (Fig. 5.54)
Stroke	67.0 mm
Bore	72.0 mm
Displacement	1091 cm^3
Compression ratio	9.0
Power	71 hp at 6000 rpm
Torque	92 Nm at 4300 rpm
Specific power	65.1 hp/litre
Valvetrain	2 overhead valves per cylinder, operated by rocker arms driven by two overhead camshafts (Fig. 5.54)
Fuel feeding	Two twin-barrel petrol Solex C32 carburettors
Ignition	Single spark plug per cylinder supplied by an inductive discharge system with a high-voltage Magneti Marelli distributor
Cooling	Liquid-fed by a centrifugal pump
Lubrication	Forced by a gear pump
Application	Lancia Fulvia 2C (sedan)

Fig. 5.53 The automotive engine Lancia 818.100–*Fulvia 2C*, 1964–1969. Kind donation of Mr. Giorgio Onori

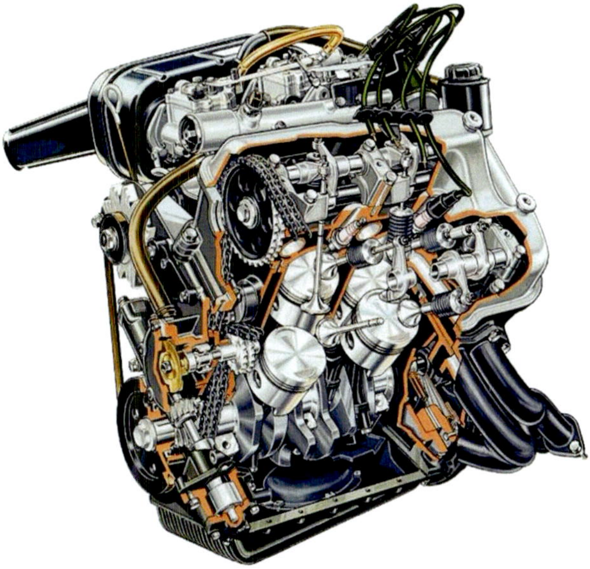

Fig. 5.54 The automotive engine Lancia 818.100–*Fulvia 2C*, 1964–1969

5.21 FIAT 135B–*Dino 2000*

1966–1969

This engine 135B (Fig. 5.55) represents the result of the FIAT-Ferrari collaboration in the mid-1960's, originally conceived to quickly make a new engine compliant with 1967 F2 Championship rules, prescribing the use of engine derived from car produced in at least 500 units. To this purpose FIAT designed a new sport car, the Dino spider, powered by a 2 litres V6 engine (Table 5.21), strictly developed starting from the Ferrari competition engine Dino 206. To this purpose, the original project was deeply revised to meet road car and mass production requirements by Aurelio Lampredi, who was one of the most important Ferrari and then FIAT mechanical designer between the 50's and 60's. As a result, the FIAT Dino spider reached a good success and was produced in over 500 examples. Thanks to its high performance, the engine was also used for the FIAT Dino coupé and later, in 1968, powered the sport car SEFAC Dino 206 GT, which was the first mid-rear engine road car (*berlinetta*) made by "Ferrari".

Table 5.21 Technical specifications of the FIAT 135B–*Dino 2000* powertrain

Manufacturer	FIAT, Turin, Italy, from 1966 to 1969
Type	135B.000
Description	V-65° 6-cylinder engine, cylinder block, heads and oil pan made of aluminium alloy with steel cylinder liners (Fig. 5.56)
Stroke	57 mm
Bore	86 mm
Displacement	1987 cm^3
Compression ratio	9.0
Power	160 hp at 7200 rpm
Torque	172 Nm at 6000 rpm
Specific power	80.5 hp/litre
Valvetrain	2 overhead valves per cylinder, operated by two overhead camshafts in each head driven by double roller chains and gears
Fuel feeding	Three twin-barrel petrol carburettor Weber 40 DCN
Ignition	Single spark plug per cylinder supplied by a Magneti Marelli distributor, controlled by a Dinoplex capacitor discharge ignition system
Cooling	Liquid-fed by a centrifugal pump
Lubrication	Forced by a gear pump
Clutch	Single-disc type
Transmission	FIAT 135 AC 100 manual gearbox, 5 speed + reverse
Application	FIAT Dino 2000 coupé and spider (Fig. 5.57); SEFAC Dino 206 GT (2-seater coupé, 1968-1969) with different gearbox and powertrain layout

Fig. 5.55 The automotive engine FIAT 135B with a FIAT 5-speed gearbox, 1966–1969

Fig. 5.56 Transparent view of the FIAT 135B engine with a FIAT 5-speed gearbox

Fig. 5.57 The FIAT Dino 2000 spider powered by the FIAT 135B engine, 1966–1969

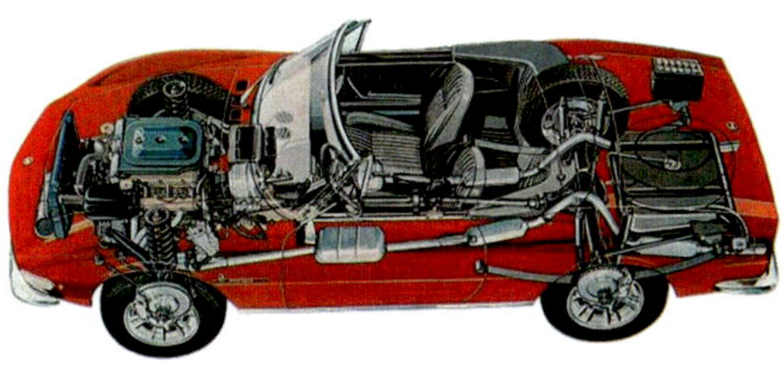

5.22 FIAT 135C–*Dino 2400*

1969–1976

The FIAT 135C (Fig. 5.58) engine has been introduced in 1969, replacing the previous 2000 cm3 model (Fig. 5.55). The new engine featured a robust cast-iron cylinder block with a better reliability and a greater power output of +20 hp (Table 5.22 and previous Table 5.21 for a technical comparison). Thanks to its high performance, it has been used for the renewed FIAT Dino coupè and spider as well as for the new SEFAC (Ferrari) Dino 246 GT/GTS. Starting from 1973 the engine has been also used for the Lancia Stratos sport car, which reached numerous successes in several international competitions, winning the Group 4 world rally championship in 1974, 1975 and 1976.

Table 5.22 Technical specifications of the FIAT 135C–*Dino 2400* powertrain

Manufacturer	FIAT, Turin, Italy, from 1969 to 1976
Type	135C.000
Description	V-65° 6-cylinder engine, cast-iron cylinder block, heads and oil pan made of aluminium alloy (Fig. 5.59)
Stroke	60 mm
Bore	92.5 mm
Displacement	2419 cm^3
Compression ratio	9.0
Power	180 hp a 6600 rpm
Torque	216 Nm a 4600 rpm
Specific power	74.4 hp/litre
Valvetrain	2 overhead valves per cylinder, operated by two overhead camshafts in each head driven by double roller chains and gears
Fuel feeding	Three twin-barrel petrol carburettor Weber 40 DCNF
Ignition	Single spark plug per cylinder supplied by a Magneti Marelli distributor, controlled by a Dinoplex capacitor discharge ignition system
Cooling	Liquid-fed by a centrifugal pump
Lubrication	Forced by a gear pump
Clutch	Single disc operated by a hydraulic master-slave system
Transmission	ZF S5-18/3 manual gearbox, 5 speed + reverse
Application	FIAT Dino 2400 (coupé and spider, 1969-1972); with different gearbox and powertrain layout: SEFAC Dino 246 GT/GTS and 246 GTS (1969-1973) and Lancia Stratos (2-seater coupé, 1974-1976)

Fig. 5.58 The automotive engine FIAT 135C with a ZF 5-speed gearbox, 1969–1976

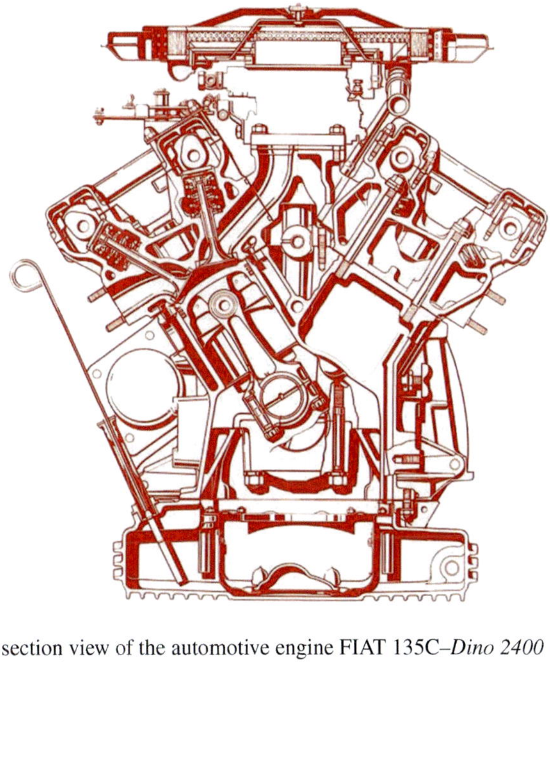

Fig. 5.59 Cross section view of the automotive engine FIAT 135C–*Dino 2400*

5.23 Lancia 831 A7–*1600 Turbo*

1983–1986

Originally developed for aircraft engine, turbocharging systems has been employed in automotive from the 1970s both for racing and sportier road car. In 1983 Lancia launched the first turbocharged version of its Delta, named HF turbo, powered by a 130 hp engine. The new 831 A7 (Fig. 5.60) was the first FIAT-Lancia mass-produced engine to be endowed with a turbo unit driven by exhaust gas, following previous models equipped with Roots compressors. The engine was derived from twin-cam four-cylinder model designed by Aurelio Lampredi and widely used in many FIAT and Lancia models. The turbocharging system required various modifications, including a reduced compression ratio, from 9.3 to 8.1, sodium-cooled exhaust valves, a lubricant oil radiator and a compressed air cooler (Fig. 5.61). The engine was equipped with a new electronically controlled ignition system created by Magneti Marelli and further endowed with a knock sensor to detect and counteract anomalous combustion phenomena at higher engine loads. A Garrett T3 unit was used to obtain a maximum intake manifold absolute pressure of 1.9 bar, automatically limited by a wastegate valve for the by-pass regulation of the exhaust gas. The first version of the engine was equipped with a double-barrel carburettor placed downstream of the compressor (Table 5.23), which was replaced with an electronic controlled port fuel injection from 1986, as well as in all successive FIAT-Lancia turbocharged engines.

Table 5.23 Technical specifications of the Lancia 831 A7–*1600 Turbo*

Manufacturer	Lancia, Turin, Italy, from 1983 to 1986
Type	831 A7.000
Description	Inline 4-cylinder engine, aluminium alloy head, cast-iron cylinder block, crankshaft with 5 main bearings
Stroke	71.5 mm
Bore	84 mm
Displacement	1585 cm^3
Compression ratio	8.0
Power	130 hp at 5600 rpm
Specific power	82.0 hp/litre
Valvetrain	2 overhead valves per cylinder, operated by two overhead camshaft driven by single toothed belt
Fuel feeding	Twin-barrel petrol carburettor Weber 32 DAT 18/250, placed downstream of the compressor
Supercharging	Exhaust gas turbocharger Garrett T3 with intercooler
Ignition	Single spark plug per cylinder supplied by an electronically controlled inductive discharge system Magneti Marelli IAW
Cooling	Liquid-fed by a centrifugal pump
Lubrication	Forced by a gear pump with oil cooler
Application	Lancia Delta HF Turbo

Fig. 5.60 The automotive engine Lancia 831 A7–*1600 Turbo*, 1983–1986

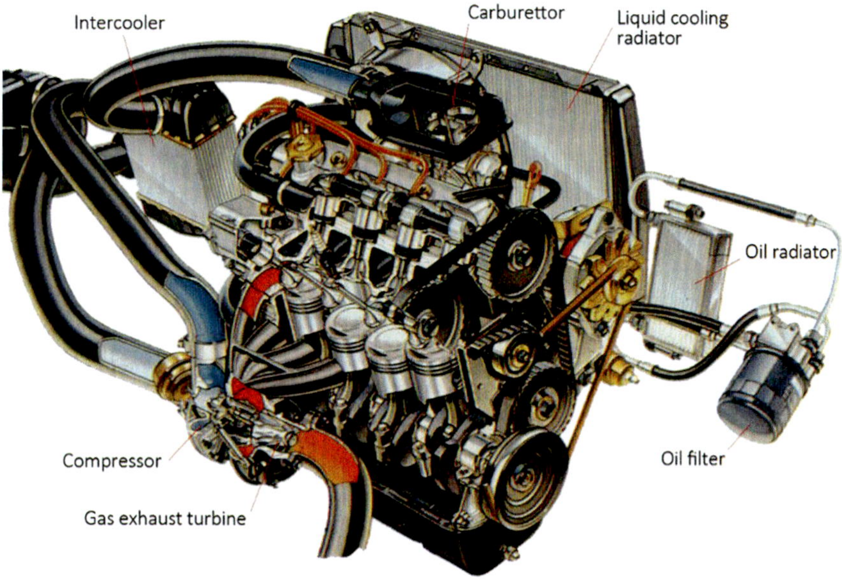

Fig. 5.61 Transparent view of the automotive engine Lancia 831 A7–*1600 Turbo*

5.24 Ducati Monster 600

1994–2001

In 1993 Ducati introduced the first model of the famous naked motorcycle named Monster equipped with a 900 cm3 V-90° 2-cylinder engine, featuring the iconic desmodromic valvetrain system. In 1994 the model-range was enlarged with the introduction of both 400 cm3 and 600 cm3 versions (Figs. 5.62 and 5.63). The latter, produced up to 2001, was the best seller version. All series featured an extensive use of aluminium alloy with a powertrain arrangement composed of engine and gearbox integrated in the same case (Table 5.24).

Fig. 5.62 The motorcycle Ducati Monster 600, 1994–2001

Table 5.24 Technical specifications of the Ducati Monster 600 powertrain

Manufacturer	Ducati, Borgo Panigale (Bologna) Italy, 1994–2001
Type	Monster 600
Description	V-90° 2-cylinder engine, gearbox-engine crankcase, cylinders and heads made of aluminium alloy
Stroke	58 mm
Bore	80 mm
Displacement	583 cm^3
Compression ratio	10.7
Power	51 hp at 8000 rpm
Torque	50 Nm at 6000 rpm
Valvetrain	2 overhead valves per cylinder, operated by a desmodromic system (Fig. 5.64)
Fuel feeding	Two single-barrel petrol Mikuni carburettors
Ignition	Single spark plug in each cylinder electronically controlled by an inductive discharge system
Cooling	Air-fed
Lubrication	Forced by a mechanical pump
Clutch	oil-wet multiple-plate type
Transmission	5-speed sequential manual gearbox
Application	Ducati Monster 600 (Fig. 5.62)

Fig. 5.63 The motorcycle engine Ducati Monster 600, 1994–2001

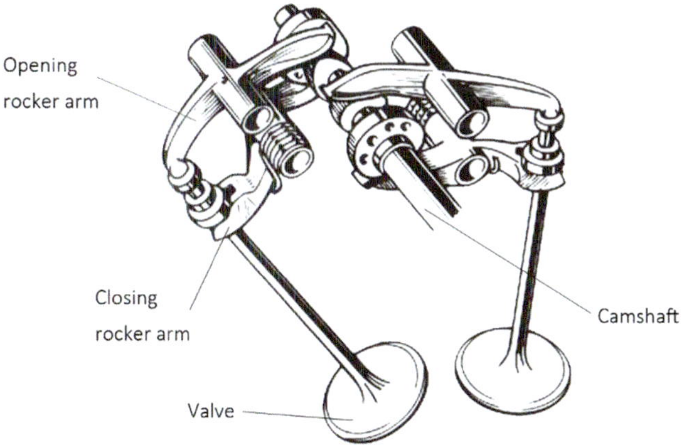

Fig. 5.64 The Ducati desmodromic valvetrain

5.25 Alfa Romeo AR33401–*Boxer 1.7 16V IE*

1994–1996

The first Alfa Romeo 4-cylinder boxer engine was originally designed in the mid-1960s for the new small family car named Alfasud. Over the years, the engine underwent several developments, such as larger displacements, fuel system upgrades from an original single carburettor up to an electronically controlled injection and ignition system. These improvements allowed the successfully use of boxer engines in both the first and second series of the Alfa Romeo 33 for over ten years. In 1990 a 133 hp engine with new 4-valve per cylinder heads was introduced for the sportiest models of the Alfa Romeo 33. From this model was finally obtained the latest boxer version, the AR33401 (Figs. 5.65 and 5.66), equipped with an upgraded new electronic control system with exhaust gas treatment, which powered the Alfa Romeo 145 and 146 1.7 16V from 1994 to 1996 (Table 5.25).

Table 5.25 Technical specifications of the Alfa Romeo AR33401—*Boxer 1.7 16V IE*

Manufacturer	Alfa Romeo, Milan, Italy, from 1994 to 1996
Type	AR33401
Description	Boxer 4-cylinder engine, heads and pistons made of aluminium alloy, cast iron cylinder block, crankshaft with 3 main bearings (Fig. 5.66)
Stroke	72.2 mm
Bore	87 mm
Displacement	1712 cm^3
Compression ratio	10.0
Power	133 hp at 6500 rpm
Torque	157 Nm at 4600 rpm
Specific power	77.7 hp/litre
Valvetrain	4 valves per cylinder with hydraulic tappets, controlled by two overhead camshafts in each head driven by toothed belts
Fuel feeding	Multipoint port fuel (petrol) injection controlled by an electronic Bosch Motronic M2.10.3, with single throttle valve per cylinder (Fig. 5.66)
Ignition	Single spark plug per cylinder, controlled by an integrated (ignition-injection) electronic system
Cooling	Liquid-fed by a centrifugal pump
Lubrication	Forced by a gear pump
Clutch	Single-disc operated by a hydraulic master-slave system
Transmission	5-speed and reverse manual gearbox
Application	Alfa Romeo 145 1.7 16V and 146 1.7 16V

Fig. 5.65 The automotive engine Alfa Romeo AR33401 with its 5-speed gearbox, 1994–1996. Kind donation of Mr. Aldo Arnetta

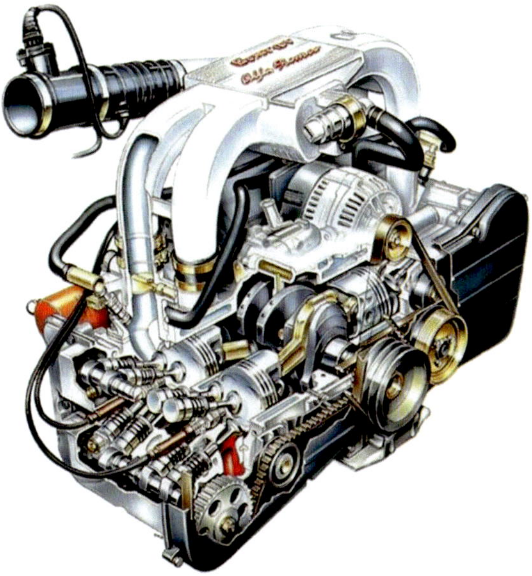

Fig. 5.66 Transparent view of the 4-cylinder boxer engine Alfa Rome AR33401

5.26 Alfa Romeo AR32401–*2.5 V6 24V*

1997–2005

At the end of the 1960s, Alfa Romeo began the development of a new V-60° six-cylinder engine conceived for its medium and high-class sedans, in place of the previous 6-inline model. After a slowdown of the project due to the consequences of the 1973 Oil Crisis, in 1979 the first 2.5 litres version of the engine made its debut, equipping the Alfa 6 sedan and then the GTV 6 coupé. In the following years the engine was developed thanks to various technical and performance improvements, resulting in new models ranging from two to three litres, including a supercharged 2-litre version. These engines were used for a long time in various cars, including the Alfa Romeo 75, 164, 155, 166 and 156, and other Lancia and FIAT sedans, as well as in sport competitions by means of appropriate modifications. In 1992 the V6 engines was renewed with new 4 valves per cylinder heads, initially introduced in the 3-litre model. From this was derived the 2.5 litre 190 hp version for the Alfa Romeo 156 sedan (Figs. 5.67 and 5.68), presented in 1997, and later also used in

Table 5.26 Technical specifications of the Alfa Romeo AR32401–*2.5 V6 24V*

Manufacturer	Alfa Romeo, Arese, Milan, from 1997 to 2005
Type	AR32401
Description	V-60° 6-cylinder engine, open deck crankcase made of aluminium alloy, crankshaft with 4 main bearings, heads and pistons made of aluminium alloy (Fig. 5.68)
Stroke	68.3 mm
Bore	88 mm
Displacement	2492 cm^3
Compression ratio	10.3
Power	190 hp at 6300 rpm
Torque	222 Nm at 5000 rpm
Valvetrain	4 valves per cylinder, controlled by two overhead camshafts in each bank, driven by a single toothed belt
Feeding	Bosch Motronic ME 2.1 multipoint electronic petrol injection system, with an electronically controlled throttle valve
Ignition	Single spark plug per cylinder, with electronic control system integrated with the fuel system management
Cooling	Liquid-fed by a centrifugal pump
Lubrication	Forced by a gear pump
Anti-pollution	Trivalent catalytic converter with oxygen sensor, polluting emission level Euro 2, later Euro 3
Application	Alfa Romeo 156 and 166

Fig. 5.67 The automotive engine Alfa Romeo AR32401–*2.5 V6 24V*, also called "Busso" as all 1979-2005 60°-V Alfa Romeo engine, from the surname of one of their most famous designers. The engine on display is equipped with its original 6-gear and reverse manual gearbox, partially sectioned for educational/promotional purposes

Fig. 5.68 Transparent view of the automotive engine Alfa Romeo AR32401–*2.5 V6 24V*

some versions of the higher-class sedan 166. Thanks to its technical and performance qualities (Table 5.26), in 2000 the 2.5 litre engine received the prestigious international "Engine of the year award" for the 2.0-2.5 litres category.

5.27 Ferrari F136 E

2004–2009

The engine of the Ferrari F430, named Tipo F136 E (Fig. 5.69), is a 90° V8 engine, designed according to the typical Ferrari flat-plane crankshaft arrangement. The experience gained by Ferrari in Formula 1 allowed significant improvements of the engine, compared to its predecessor, in terms of performance and weight, with a reduction in overall dimensions. Despite a 20% increase in displacement, the weight of the engine has increased by only 4 kg, while performance has significantly improved with a maximum power of 490 hp at 8500 rpm. The engine is extremely compact thanks to a dry sump lubrication system, integrated into the crankcase. Thanks to a deep development of the internal fluid dynamics, the combustion efficiency was optimized, achieving an impressive specific power of 114 hp/litre in compliance with the highest emission standards in force at the time of the car's launch: Euro 4 and LEV 2 (Table 5.27).

Table 5.27 Technical specifications of the Ferrari F136 E

Manufacturer	Ferrari S.p.A., Maranello, from 2004 to 2009
Type	F136 E
Description	V-90° 8-cylinders engine, crankcase, heads and pistons made of aluminium alloy, flat-plane crankshaft with 5 main bearings
Stroke	81 mm
Bore	92 mm
Displacement	4308 cm^3
Compression ratio	11.3
Max. power	490 hp at 8500 rpm
Max. torque	465 Nm at 5250 rpm (BMEP 13.6 bar)
Mean piston speed	23 m/s at 8500 rpm
Electronic control system	Bosch Motronic ME7 with two control units, two motorized throttles and active anti-knock control
Valvetrain	4 valves per cylinder, controlled by two camshafts in each bank driven by chains, with inlet and exhaust variable valve timing system
Fuel feeding	Electronically controlled petrol indirect injection, with variable geometry intake ducts
Ignition	Electronically controlled with one spark plug per cylinder, powered by an independent coil per cylinder
Cooling	Liquid-fed by a centrifugal pump
Lubrication	Dry-sump type, forced by gear pumps with an oil-water heat exchanger
Mass	175 kg
Power-to-mass ratio	2.8 hp/kg
Application	Ferrari F430 (Fig. 5.70)

Fig. 5.69 The automotive engine Ferrari F136 E, 2004–2009. Kind donation of Ferrari S.p.A

Fig. 5.70 The Ferrari F430 powered by the F136 E engine allowing to reach 100 km/h from standstill in 4 s, 200 km/h in 12.2 s and a top speed of over 315 km/h

5.28 Subaru EJ257–*2.5 Turbo*

2004–2012

Starting in 1966, with the introduction of its first boxer engine, Subaru developed a vast series of models ranging from 4 to 6 cylinders. This arrangement, together with all-wheel drive, gained an international success becoming two of the characterizing features of Subaru's technical design. The 300 hp EJ257 model is one of the most powerful (122 hp/litre) Subaru mass-produced engines (Table 5.28), resulted from a detailed development which included turbocharging and variable valve timing system. It has been produced since 2004 for the Impreza WRX STi (Fig. 5.71), that achieved (by competition versions) numerous successes in international rally championships.

Fig. 5.71 The Subaru Impreza WRX STi powered by the engine EJ257–*2.5 Turbo*

Table 5.28 Technical specifications of the Subaru EJ257–*2.5 Turbo*

Manufacturer	Subaru, Fuji Heavy Industries, Japan, 2004–2012
Type	EJ257 (Fig. 5.72)
Description	Boxer 4-cylinder engine, cylinder block, heads and pistons made of aluminium alloy (Fig. 5.73)
Stroke	79 mm
Bore	99.5 mm
Displacement	2457 cm^3
Compression ratio	8.71
Power	300 hp at 6000 rpm
Torque	407 Nm at 4000 rpm
Valvetrain	4 valves per cylinder with hydraulic tappets, controlled by two overhead camshafts in each head driven by toothed belts, intake variable timing system
Fuel feeding	Multipoint port fuel (petrol) injection operated by an electronic control system integrated with ignition
Supercharging	Exhaust gas turbocharger IHI with and air-to-air cooler
Ignition	Single spark plug and ignition coil per cylinder
Cooling	Liquid-fed by a centrifugal pump
Lubrication	Forced by a gear pump
Mass	120 kg
Application	Subaru Impreza WRX STi (Fig. 5.71)

Fig. 5.72 The automotive boxer engine Subaru EJ257–2.5 *Turbo*, 2004–2012. Kind donation of Mari Service s.r.l

Fig. 5.73 Transparent view of the boxer engine Subaru EJ257–2.5 *Turbo*, 2004–2012

5.29 Ford EcoBoost 1.0 L

2012

The EcoBoost 1.0 L (Fig. 5.74) engine was developed at the Ford Research Center in Dunton (UK) in cooperation with the FEV research centre in Germany. From 2012 it has been produced for various Ford models. Its design shows great attention for engine efficiency and lightweight achieved by means of many fine technical features: downsizing with 3-inline-cylinder layout, direct in-cylinder fuel injection, inlet and exhaust variable valve timing, low inertia turbocharger and variable displacement oil pump (Table 5.29). As a result, from 2012 to 2017, the engine gained the prestigious international award "Engine of the year" in the less than one-litre category.

Table 5.29 Technical specifications of the Ford EcoBoost 1.0 L

Manufacturer	Ford Motor Company, from 2012
Type	EcoBoost 1.0 L SCTi GTDi
Description	Inline 3-cylinder engine, cast-iron cylinder block, head and pistons made of aluminium alloy, special cast-iron alloy crankshaft with 4 main bearings
Stroke	81.9 mm
Bore	71.9 mm
Displacement	998 cm^3
Compression ratio	10.0
Power	125 hp at 6000 rpm
Torque	170 Nm between 1400 to 4500 rpm (Fig. 5.75) 200 Nm with over boost
Specific power	125.3 hp/litre
Electronic control system	Bosch MED17 with CAN-Bus system, including knock detection for each cylinder
Valvetrain	4 valves per cylinder with hydraulic tappets, controlled by two overhead camshafts, with intake and exhaust variable timing system, driven by an oil-wet toothed belt
Fuel feeding	High pressure petrol direct in-cylinder injection
Supercharging	Exhaust gas turbocharger made by Continental, max. impeller speed 248000 rpm, max. turbine exhaust gas inlet temperature 1050 °C, air-to-air intercooler
Ignition	Single spark plug and single ignition coil per cylinder
Cooling	liquid-fed by a centrifugal pump, with 2 independent thermostatic valve/circuit, for faster head warm-up
Lubrication	Variable displacement pump and oil cooler
Mass	97 kg (dry)
Exhaust aftertreatment system	Three-way catalytic converter (Euro 5)
Application	Fiesta (incl. B-max), Focus (incl. C-Max), Transit Courier

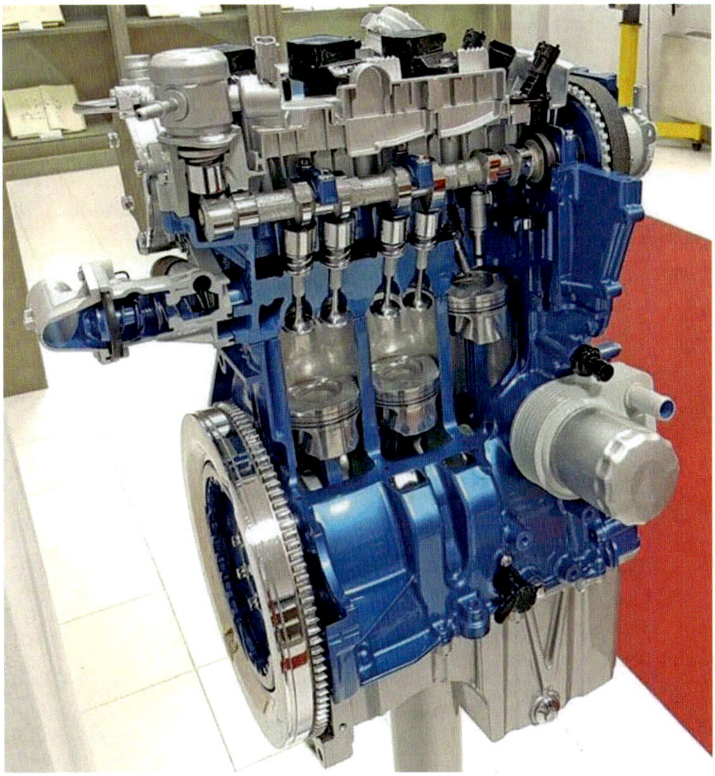

Fig. 5.74 The Ford EcoBoost 1.0 L engine, partially sectioned for educational purposes. Kind donation of Ford Italia S.p.A

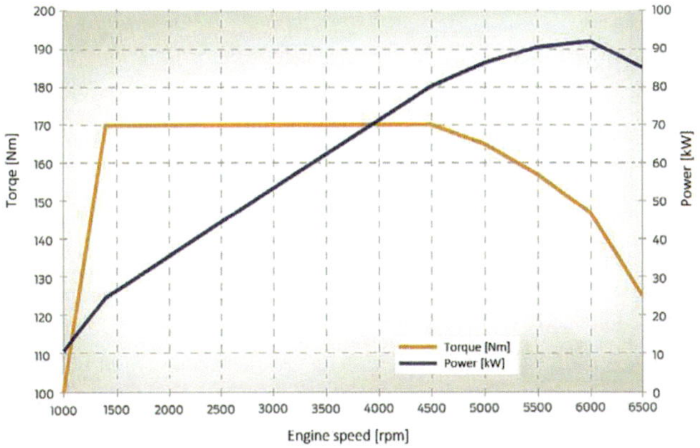

Fig. 5.75 Power and torque curves of the automotive engine Ford EcoBoost 1.0 L

5.30 Lamborghini L539–*Aventador*

2010–2016

The L539 (Fig. 5.76) is a high-performance naturally aspirated 6.5-litre 12-cylinder engine, manufactured by Lamborghini Automobili for its sport car Aventador. At its debut in 2010, the engine represented the top-level production of Lamborghini as well as one of the most powerful engines for commercial sport car, thanks to its maximum power output of 700 hp, with a mass-to-power ratio of 2.98 hp/kg (Table 5.30). It has been developed from the previous generation V-12 L537 engine, by redesigning the main components and introducing several technical improvements such as a new combustion chamber design and a new electronic control system. This was developed by Lamborghini in collaboration with EFI Technology of Bologna, and it also includes the fully integrated management of the four-wheel drive transmission with a 7-speed robotised gearbox. This powertrain allows the Lamborghini Aventador to reach 100 km/h from standstill in 2.9 seconds and 200 km/h in 8.6 seconds.

Table 5.30 Technical specifications of the Lamborghini L539–*Aventador*

Manufacturer	Automobili Lamborghini S.p.A., Sant'Agata Bolognese, Italy, from 2010 to 2016
Type	L539
Description	V 60° 12-cylinder engine, open-deck cylinder block, heads and pistons made of aluminium alloy
Stroke	76.4 mm
Bore	95.0 mm
Displacement	6498 cm^3
Compression ratio	11.8
Power	700 hp at 8250 rpm (max. speed 8500 rpm)
Torque	690 Nm at 5500 rpm (Fig. 5.77)
Mean piston speed	21 m/s at 8250 rpm
Specific power	107.7 hp/litre
Electronic control	Composed of two control units (master-slave) for the entire powertrain management
Valvetrain	4 valves per cylinder operated double overhead camshafts in each bank with inlet and exhaust variable valve timing system
Fuel feeding	Port fuel (petrol) injection, with variable geometry air inlet ducts
Ignition	Single spark plug with a dedicated ignition coil in each cylinder; spark plugs are also used for knock and misfire detection by a ion-sensing-based system
Cooling	Liquid-fed by a centrifugal pump
Lubrication	Dry-sump type, composed of 8 pumps, with 2 radiators
Mass	235 kg
Application	Lamborghini Aventador LP 700-4 (Fig. 5.78)

Fig. 5.76 The V-12-cyinder L539 engine of the Lamborghini Aventador LP 700-4, 2010–2016. Kind donation of Lamborghini Automobili S.p.A

Fig. 5.77 Power and torque curves of the Lamborghini L539 engine

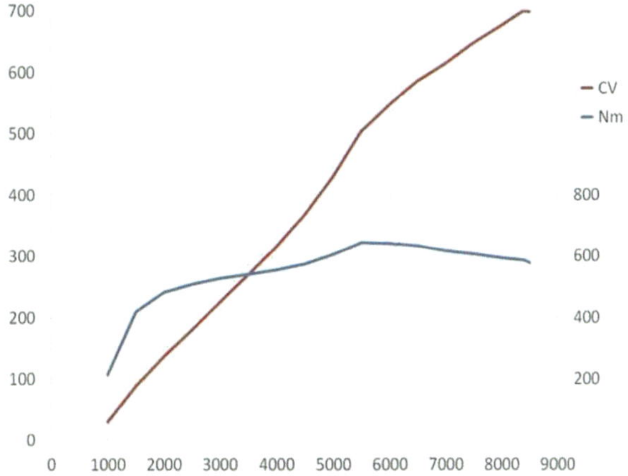

Fig. 5.78 The Lamborghini Aventador LP 700-4 powered by the 12-cylinder engine L 539

5.31 Harley-Davidson Milwaukee Eight 117

2018

The Milwaukee Eight, presented in 2016, represents the ninth generation of Harley-Davidson big twin-cylinder engines, but from a purely technical point of view it should only be considered the third completely new project in 80 years of activity of the most famous US motorcycle company, which is particularly faithful to its technical traditions. The classic V-45° twin-cylinder arrangement is unchanged, but the new model has a more robust structure and is equipped with a counterbalance shaft driven by the crankshaft to reduce the engine vibrations. The name of the engine derives from the city where the company is based followed by the total number of valves. Respect to the previous model, the valvetrain is operated by a single camshaft in the crankcase and each cylinder head is equipped with two spark plugs. Together with other technical upgrades, these features allowed a reduction in fuel consumption and a 10% torque increase with the same engine mass. Available in different configurations and various technical upgrades, the model 117 (i.e. 1.9 litres)

Table 5.31 Technical specifications of the Harley-Davidson Milwaukee Eight 117

Manufacturer	Harley-Davidson, Milwaukee, Winsconsin, USA, since 2018
Type	Milwaukee Eight 117 (Fig. 5.79)
Description	Four-stroke motorcycle engine with 2 cylinders in a 45° V, aluminum alloy cylinder block with two main supports for roller bearings, and counter-rotating balancer shaft, aluminum alloy heads, pistons and cylinders with steel liners (Fig. 5.80)
Stroke	114.3 mm
Bore	103.5 mm
Displacement	1923 cm^3
Compression ratio	10.2
Power	107 hp at 4750 rpm
Torque	170 Nm at 3750 rpm
Specific power	55.6 hp/litre
Valvetrain	2 overhead valves per cylinder, controlled by pushrods and rocker arms with single camshaft in the crankcase
Fuel feeding	Port fuel (petrol) injection operated by an electronic control system (integrated with ignition management)
Ignition	Two spark plugs and a knock sensor in each cylinder
Cooling	Operated by the lubricant oil, with an additional liquid-fed circuit (optional) for heads cooling improvement
Lubrication	Forced by a gear pump and oil cooler
Application	Harley-Davidson Street Glide ST and Road Glide ST Low Rider S and ST, including special models CVO

is dedicated to the top-series models of Harley-Davidson motorbikes (Fig. 5.81 and Table 5.31).

Fig. 5.79 The motorcycle engine Harley-Davidson Milwaukee Eight 117, 2018. Kind donation of Mr. Gilberto Bonafede

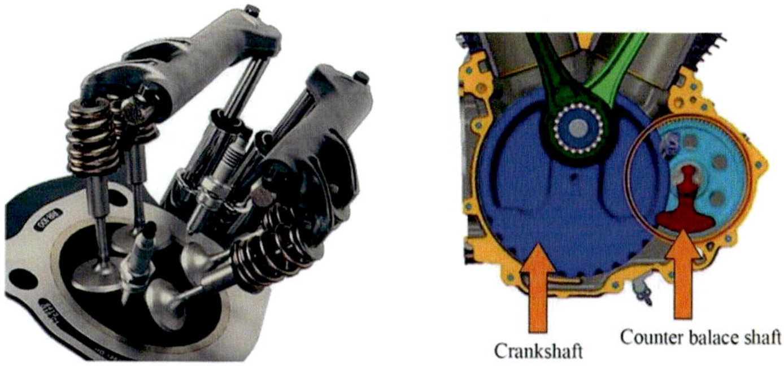

Fig. 5.80 The 4-valve twin-spark plug engine head (left) and the counterbalance shaft (right)

Fig. 5.81 A Harley-Davidson Street Glide ST powered by the Milwaukee Eight 117

5.32 Toyota 2ZR-FXE–*1.8 Hybrid E-CVT*

2018–2023

In 1990s Toyota was one of the main manufacturers which invested considerable resources modern hybrid propulsion systems and the first to launch a hybrid production car, the Prius, at the end of 1997. Since that model it achieved a long and consolidated tradition developing various types of hybrid systems. The ZR series, introduced in 2007, includes various 4-cylinder models with different power outputs. The latest versions of these engines are thermal units of hybrid powertrains featuring a high-level efficiency reaching up to 40% in optimal operative conditions. These performances, at the top of their category, are obtained thanks to various advanced features both in the internal combustion engine and in the electric units (Fig. 5.82): such as variable valve timing system, which allows the engine to operate according to the Miller-Atkinson cycle, and Power-Split transmission, having continuously variable speed ratio. This combines the advantages of the more traditional series-hybrid and parallel-hybrid systems, eliminating their respective disadvantages: the power-split unit, whose fundamental element is an epicyclic gear train, allows the decoupling of the internal combustion engine from the road load as in the hybrid-series system, as well as, the simultaneous use of the engine together with all electric motors of the powertrain as in the parallel-hybrid arrangement (Table 5.32).

Fig. 5.82 The hybrid powertrain Toyota 2ZR-FXE–*1.8 Hybrid E-CVT*, 2018–2023. Kind donation of Toyota Motor Italia S.pA

Table 5.32 Technical specifications of the Toyota 2ZR-FXE–*1.8 Hybrid E-CVT*

Manufacturer	Toyota Motor Corporation, Toyota (Aichi), Japan, 2016 to 2022
Type	2ZR-FXE with P610 transmission (Fig. 5.82)
Overall description	integrated hybrid powertrain, with internal combustion engine and two electric motor-generators
Internal combustion engine	4 in-line cylinders, head and cylinder block made of aluminium alloy, with 8 mm off-center crankshaft
Stroke	88.3 mm
Bore	80.5 mm
Displacement	1798 cm^3
Compression ratio	13.0
Max. power	98 hp at 5200 rpm
Max. torque	142 Nm at 3600 rpm
Electronic control	Integrated management for engines and transmission
Valvetrain	4 valves per cylinder controlled, via roller rocker arms, by two overhead camshafts driven by an oil-wet toothed belt; intake variable valve timing (VVT) for Miller-Atkinson cycle in engine part load conditions
Fuel feeding	Electronically controlled port fuel (petrol) injection, variable geometry air inlet ducts
Ignition	Single spark plug-coil per cylinder supplied by an electronically controlled inductive discharge system
Cooling	Liquid, with electronically controlled electric pump and exhaust gas heat recovery (EHR)
Lubrication	Forced by a gear pump with automatic by-pass valve for pressure and flow rate control according to the engine speed
Anti-pollution	Trivalent catalytic converter, exhaust gas recirculation (EGR) system and particulate filter; emissions level Euro 6d
Electric propulsion	P610 (E-CVT), 2 synchronous motor-generators brush-less with permanent magnets
Max. power MG1 engine	6.8 hp, liquid cooled
Max. power MG2 engine	72 hp at 600 V, max. torque 163 Nm, liquid cooled
Max. power combined	122 hp (by engine and electric motors)
Transmission	Aichi P610 continuously variable ratio power-split transmission, consisting of an epicyclic gearing connected to the engine and the two electric motor-generators, with various series-parallel driving configurations between the three drive units
Total mass	150 kg
Total power-to-mass ratio	0.81 hp/kg
Application	Toyota Prius and Corolla 1.8 Hybrid, Lexus CT et al.

Chapter 6
Didactic Mechanisms and Models

Abstract The Museum displays an extensive array of educational equipment, whose older specimens date back to the Royal School of Application for Engineers and Architects, founded in 1866. Over a hundred metal and wooden scale models of elementary machines and kinematic systems illustrate the working principles of a large number of mechanical systems nowadays still widely used for several applications. Due to its ancient origin a significant part is dedicated to steam engines, illustrating various control system with variable valve timing later reused in many other fields of industry and motoring. Many of these models was manufactured by esteemed German and Italian firms like Schröder of Darmstadt and Eisentraeger of Milan. Over time, the collection has been gradually expanded to incorporate modern and then contemporary mechanical systems, up to piston engine and turbojet models. Beyond their historical and collectible value, this heterogeneous collection of models plays a vital role in many educational activities promoted by the Museum for both academic and school students, due to their timeless efficacy on representing fundamental mechanical principles.

G. Genchi, *The Museum of Engines and Mechanisms of the University of Palermo*,
History of Mechanism and Machine Science 48,
https://doi.org/10.1007/978-3-031-63042-2_6

Fig. 6.1 The exhibit area of the Museum dedicated to didactic mechanisms and models

The Museum has a large collection of educational mechanisms, the oldest of which date back to the second half of the nineteenth century (Fig. 6.1). Many of these are part of the first teaching equipment of the Cabinet of Applied Mechanics and Machinery at the Royal School of Application for Engineers and Architects of Palermo. The collection includes over one hundred metal or wooden models of elementary machines, kinematic systems, transmission parts and mechanisms for describing coupling profiles between mechanical elements (from Figs. 6.2 to 6.42). An entire set of models is dedicated to steam engines, widely used in the nineteenth century in various sectors of civil and industrial engineering, which represent the main types valvetrain and regulation systems in reciprocating engines (from Figs. 6.43 to 6.53). Most of these models were built for educational purposes by specialized German and Italian firms, including Schröder of Darmstadt (Germany) and Eisentraeger of Milan. Over time, this collection has been progressively updated through the acquisition of didactic mechanisms created to illustrate the newly developed mechanical systems, including modern transmission systems, some hydraulic machines and, above all, the various types of small fully working steam and internal combustion engines. The latter are small full working aeromodel engines. These include steam, spark-ignition, and compression-ignition engines, both two-stroke and four-stroke, as well as two jet engine models. The oldest specimens date back to the beginning of the 1900s and are unique pieces of handcrafted construction (Fig. 6.54 and 6.55), while others were produced by specialized companies, Italian and foreign, such as the Elia of Turin and the Microtecnica of Bologna (Figs. 6.56, 6.57, 6.58, 6.59), the British International Model Aircraft Ltd. (Fig. 6.60), the American McCoy (Fig. 6.61) and the Japanese Enya (Fig. 6.63). Some of these engines have been donated to the Museum by enthusiasts and collectors (e.g. Figs. 6.60, 6.62 and many others reported in the Collection index). The educational models section plays an important role from the point of view of the educational activities in which the Museum is constantly engaged, since the wide variety of items (from Figs. 6.64 to 6.76) in the collection illustrates in detail the various steps in the evolution of machine mechanics, as well to provides a fairly broad overview of the possible solutions in the design of the individual mechanisms. Even today some models are carefully used as useful teaching tools due to their ability to clearly represent some of the most important principles of mechanics that have remained unchanged over time.

6.1 Didactic Model of a Single Stage Spur Gear Transmission

1870–1880s.

Fig. 6.2 Didactic model of a single stage spur gear transmission. Manufacturer: L. Schröder, Darmstadt, Germany, 1870–1880s

6.2 Didactic Model of a Spur Gear Transmission

1870–1880s.

Fig. 6.3 Didactic model of a spur gear transmission with central idle gear. Manufacturer: L. Schröder, Darmstadt, Germany, 1870–1880s

6.3 Didactic Model of Worm Gear Transmission

1870–1880s.

Fig. 6.4 Didactic model of a worm gear transmission. Manufacturer: L. Schröder, Darmstadt, Germany, 1870–1880s

6.4 Didactic Wooden Model of Worm Gear Transmission

1870–1880s.

Fig. 6.5 Didactic wooden model of a worm gear transmission, 1870–1880s

6.5 Didactic Model of a Bevel Gear Transmission

1870–1880s.

Fig. 6.6 Didactic wooden model of a bevel gear transmission, 1870–1880s

6.6 Didactic Model of Orthogonal Pin-Gear Transmission

1870–1880s.

Fig. 6.7 Didactic wooden model of orthogonal pin-gear transmission, 1870–1880s

6.7 Didactic Model of an Elliptical Gear Transmission

1870–1880s.

Fig. 6.8 Didactic wooden model of an elliptical gear transmission, with two eccentric pivoted wheels, showing cyclical variable speed ratio, 1870–1880s

6.8 Didactic Model of an Elliptical-Spur Gear Transmission

1870–1880s.

Fig. 6.9 Didactic wooden model with one elliptical and one spur gear, both eccentric pivoted, showing cyclical variable transmission ratio, 1870–1880s

6.9 Didactic Model of an Internal Gear Transmission

1870–1880s.

Fig. 6.10 Didactic wooden model of internal gear transmission, 1870–1880s

6.10 Didactic Model of a Pinion and Rack Transmission

1870–1880s.

Fig. 6.11 Didactic wooden model of a pinion and rack transmission, 1870–1880s

6.11 Didactic Model of a Grasshopper Escapement

1870–1880s.

Fig. 6.12 Didactic wooden model of a grasshopper escapement, a low-friction driving system for pendulum clocks invented by the British clockmaker John Harrison, around 1722

6.12 Didactic Freewheel Mechanism

1870–1880s.

Fig. 6.13 Didactic wooden model of a freewheel (one-way) driving mechanism, 1870–1880s

6.13 Didactic Model of a Bevel Gear Differential Mechanism

1870–1880s.

Fig. 6.14 Didactic model of a bevel gear differential mechanism. Manufacturer: L. Schröder, Darmstadt, Germany, 1870–1880s

6.14 Didactic Model of a Rotating Slider Transmission

1870–1880s.

Fig. 6.15 Didactic model with a rotating slider driven by a single-stage spur gear transmission. Manufacturer: L. Schröder, Darmstadt, Germany, 1870–1880s

6.15 Didactic Model of a Double Slider Crank Mechanism

1870–1880s.

Fig. 6.16 Didactic model of a double (cross-shape) slider crank mechanism with 1:2 speed ratio. Manufacturer: L. Schröder, Darmstadt, Germany, 1870–1880s

6.16 Didactic Model of a Cardano Joint

1870–1880s.

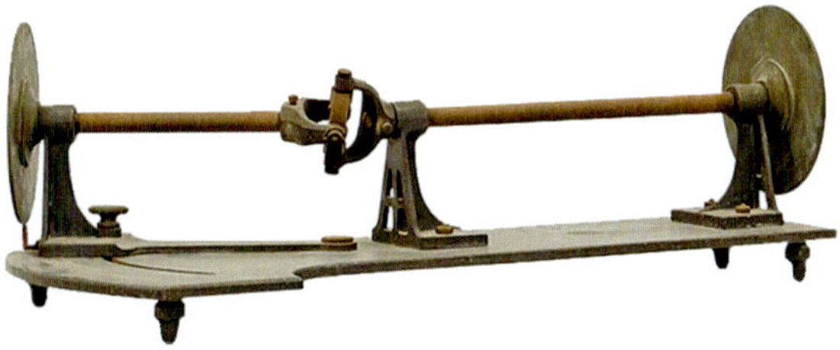

Fig. 6.17 Didactic model of Cardano joint with variable slope angle. Manufacturer: L. Schröder, Darmstadt, Germany, 1870–1880s

6.17 Didactic Wooden Model of a Cardano Joint

1870–1880s.

Fig. 6.18 Didactic wooden model of Cardano joint with variable slope angle, 1870–1880s

6.18 Didactic Model of a Reciprocating Cam-Driven Transmission

1870–1880s.

Fig. 6.19 Didactic wooden model of a reciprocating cam-driven transmission, 1870–1880s

6.19 Didactic Model of Cam with Double Internal Profile

1870–1880s.

Fig. 6.20 Didactic wooden model of cam with double internal profile, 1870–1880s

6.20 Didactic Model of a Reciprocating 4-Step Cam-Driven Transmission

1870–1880s.

Fig. 6.21 Didactic wooden model of a reciprocating 4-step cam-driven transmission, 1870–1880s

6.21 Didactic Model of a Hammer-Cam Mechanism

1870–1880s.

Fig. 6.22 Didactic wooden model of a hammer-cam mechanism, 1870–1880s

6.22 Didactic Model of a Reciprocating Cam-Driven Transmission

1870–1880s.

Fig. 6.23 Didactic wooden model of a reciprocating cam-driven transmission, 1870–1880s

6.23 Didactic Model of a Reuleaux Triangle Cam Mechanism

1870–1880s.

Fig. 6.24 Didactic model of a Reuleaux triangle cam mechanism. Manufacturer: L. Schröder, Darmstadt, Germany, 1870–1880s

6.24 Didactic Wooden Model with a Reuleaux Triangle-Cam Mechanism

1870–1880s.

Fig. 6.25 Didactic wooden model of a Reuleaux triangle cam mechanism, 1870–1880s

6.25 Didactic Model of a Cardioid Cam Mechanism

1870–1880s.

Fig. 6.26 Didactic model of a cardioid cam and roller-tappet mechanism. Manufacturer: L. Schröder, Darmstadt, Germany, 1870–1880s

6.26 Didactic Model of an Internal Cam Mechanism

1870–1880s.

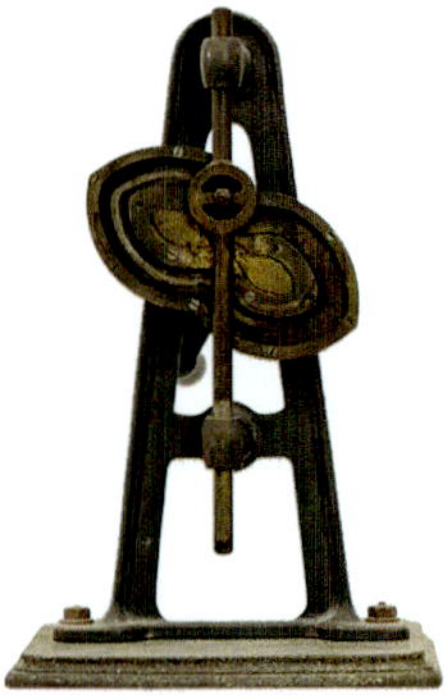

Fig. 6.27 Didactic model of an internal cam mechanism. Manufacturer: L. Schröder, Darmstadt, Germany, 1870–1880s

6.27 Didactic Model of a Chain Pulley Transmission

1870–1880s.

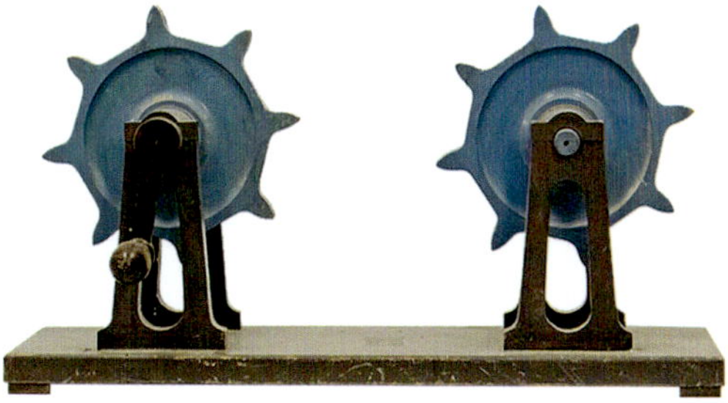

Fig. 6.28 Didactic wooden model of a chain pulley transmission (the original chain is not shown), 1870–1880s

6.28 Didactic Model of a Three-Speed Ratio Belt Pulley Transmission

1870–1880s.

Fig. 6.29 Didactic wooden model of a belt pulley transmission with three speed ratios (the original leather belt is not shown), 1870–1880s

6.29 Didactic Model of a 2-stage Belt Pulley Transmission

1870–1880s.

Fig. 6.30 Didactic wooden model of a 2-stage belt pulley transmission, 1870–1880s

6.30 Didactic Model of a Non-Parallel Axles Transmission

1870–1880s.

Fig. 6.31 Didactic model of a non-parallel axles transmission by means of roller friction elements, 1870–1880s

6.31 Didactic Model of a Continuous Variable Transmission Ratio Mechanism

1870–1880s.

Fig. 6.32 Didactic model of model of a continuous variable transmission ratio mechanism with orthogonal friction wheels, 1870–1880s

6.32 Didactic Model of Belt Pulley Transmission with Continuous Variable Ratio

1870–1880s.

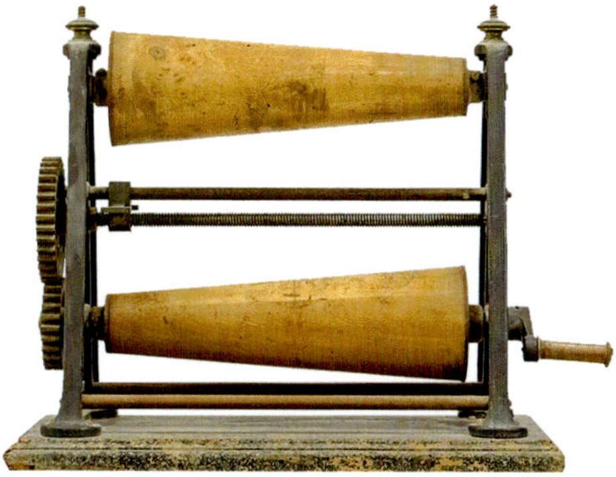

Fig. 6.33 Didactic model of a belt pulley transmission with continuous variable ratio (the original leather belt is not shown), 1870–1880s

6.33 Gyroscope

1870–1880s.

Fig. 6.34 Didactic gyroscope with counterweight, 1870–1880s

6.34 Didactic Mechanism for Rotational Dynamics Demonstration

1870–1880s.

Fig. 6.35 Didactic mechanism with asymmetric-mass oblique rotating disk for rotational dynamics demonstration. Manufacturer: L. Schröder, Darmstadt, Germany, 1870–1880s

6.35 Didactic Model of a Reverse Drive Controlled by a Watt-Type Regulator

1870–1880s.

Fig. 6.36 Didactic model of a bevel gear reverse drive controlled by a Watt-type centrifugal regulator. Manufacturer: L. Schröder, Darmstadt, Germany, 1870–1880s

6.36 Didactic Model of a Buss-Type Centrifugal Regulator

1870–1880s.

Fig. 6.37 Didactic model of a liquid/gas valve controlled by Buss-type centrifugal regulator driven by belt pulley transmission (the original leather belt is not shown), 1870–1880s

6.37 Didactic Model of a Drum Brake

1870–1880s.

Fig. 6.38 Didactic model of a drum brake operated by linkage lever. Manufacturer: L. Schröder, Darmstadt, Germany, 1870–1880s

6.38 Didactic Model of a Drum Brake with Worm Gear Actuator

1870–1880s.

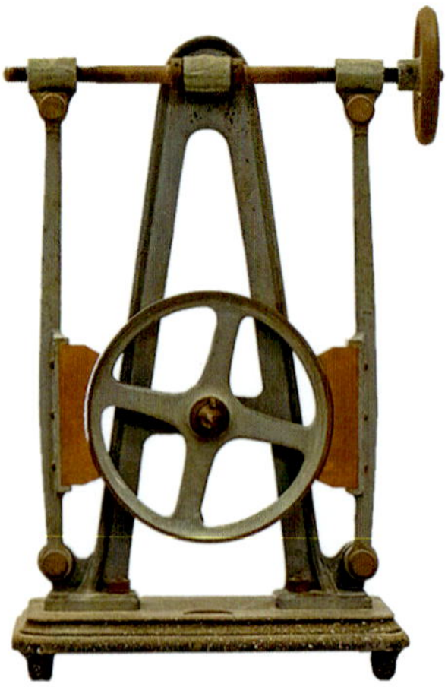

Fig. 6.39 Didactic model of a drum brake with worm gear actuator. Manufacturer: L. Schröder, Darmstadt, Germany, 1870–1880s

6.39 Didactic Mechanism of a Screw Press

1870–1880s.

Fig. 6.40 Didactic wooden model of a screw press, 1870–1880s

6.40 Didactic Model of a Balance-Beam Driving Mechanism

1870–1880s.

Fig. 6.41 Didactic model of a balance-beam driving mechanism. Manufacturer: L. Schröder, Darmstadt, Germany, 1870–1880s

6.41 Didactic Model of a Multiple Driving Balance-Beam Mechanism

1870–1880s.

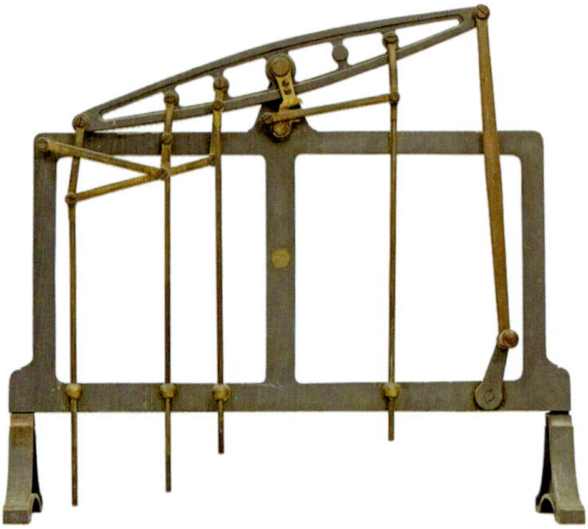

Fig. 6.42 Didactic model of a multiple driving balance-beam mechanism. Manufacturer: L. Schröder, Darmstadt, Germany, 1870–1880s

6.42 Didactic Model of a Reciprocating Steam Engine

1870–1880s.

Fig. 6.43 Didactic model of a reciprocating steam engine, with a single double-acting cylinder and adjustable valve timing. Manufacturer: L. Schröder, Darmstadt, Germany, 1870–1880s

6.43 Didactic Model of a Reciprocating Steam Engine, Stephenson-Type Variable Timing System

1870–1880s.

Fig. 6.44 Didactic model of a reciprocating steam engine, with double-acting cylinder and Stephenson-type variable timing system. Manufacturer: G. Eisentraeger, Milan, Italy, 1870–1880s

6.44 Didactic Model of a Reciprocating Steam Engine with Meyer's Slide Valve

1870–1880s.

Fig. 6.45 Didactic model of a reciprocating steam engine with Meyer's slide valve and adjustable valve timing. Manufacturer: L. Schröder, Darmstadt, Germany, 1870–1880s

6.45 Didactic Model of an Adjustable Timing System for Reciprocating Steam Engine

1870–1880s.

Fig. 6.46 Didactic model of a variable timing system for reciprocating steam engine. Manufacturer: L. Schröder, Darmstadt, Germany, 1870–1880s

6.46 Didactic Model of a Reciprocating Steam Engine with Variable Timing System and Sulzer-Type Valves

1870–1880s.

Fig. 6.47 Didactic model of a reciprocating steam engine with variable timing system and Sulzer-type valves. Manufacturer: G. Eisentraeger, Milan, Italy, 1870–1880s

6.47 Didactic Model of a Reciprocating Steam Engine with Meyer-Type Variable Timing System

1870–1880s.

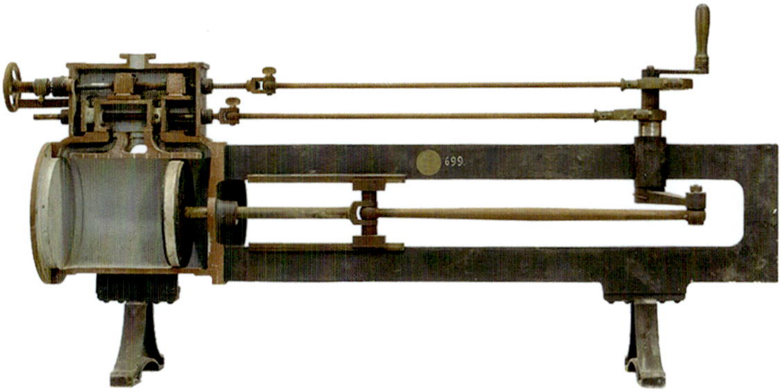

Fig. 6.48 Didactic model of a reciprocating steam engine with a Meyer-type variable timing system. Manufacturer: L. Schröder, Darmstadt, Germany, 1870–1880s

6.48 Didactic Model of a Reciprocating Steam Engine with Corliss-Type Variable Timing System

1870–1880s.

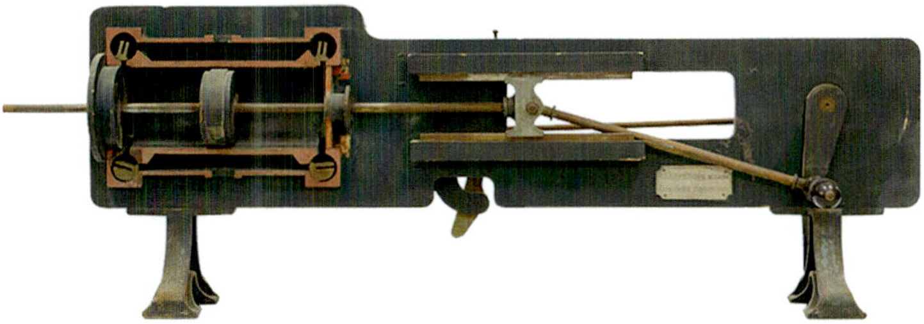

Fig. 6.49 Didactic model of a reciprocating steam engine with a Corliss-type variable timing system. Manufacturer: G. Eisentraeger, Milan, Italy, 1870–1880s

6.49 Didactic Model of a Reciprocating Steam Engine with Reuleaux Cam-Driven Adjustable Timing System

1870–1880s.

Fig. 6.50 Didactic model of a reciprocating steam engine with a Reuleaux cam-driven and adjustable timing system. Manufacturer: L. Schröder, Darmstadt, Germany, 1870–1880s

6.50 Didactic Model of a Reciprocating Steam Engine with Adjustable Timing System

1870–1880s.

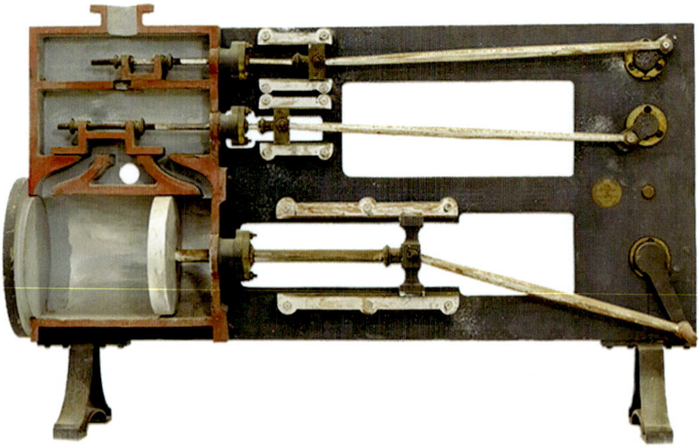

Fig. 6.51 Didactic model of a reciprocating steam engine with adjustable timing system. Manufacturer: L. Schröder, Darmstadt, Germany, 1870–1880s

6.51 Didactic Model of a Reciprocating Steam Engine with Farcot-Type Variable Timing System

1870–1880s.

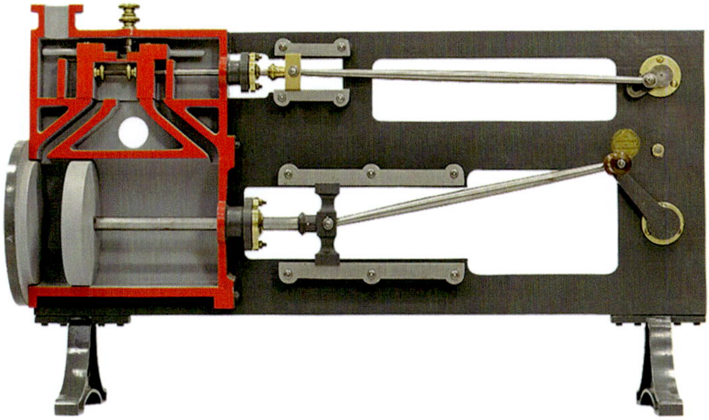

Fig. 6.52 Didactic model of a reciprocating steam engine with a Farcot-type variable timing system. Manufacturer: L. Schröder, Darmstadt, Germany, 1870–1880s

6.52 Didactic Model of a Reciprocating Steam Engine with Steam Generator

1920s.

Fig. 6.53 Didactic model of a reciprocating steam engine with steam generator, 1920s

6.53 Didactic Single-Cylinder Spark-Ignition Engine

Early '900.

Fig. 6.54 Didactic working scale model of a single-cylinder spark-ignition 4-stroke gas engine, with automatic speed regulator. Manufacturer: Antonino Fiorentino, Palermo, early '900

6.54 Didactic Scale Model of a Winch Driven by a 2-cylinder Steam Engine

1958–1959.

Fig. 6.55 Didactic working scale model of a winch driven by a 2-cylinder steam engine. Manufacturer: C.N.R. (Cantieri Navali Riuniti, part of Fincantieri S.p.A.), II Corso Aggiustatori, Palermo, Italy, 1958–1959

6.55 Elia S Aero-Model Engine

1947.

Fig. 6.56 Single-cylinder two-stroke Diesel aero-model engine with two-blade propeller. Manufacturer: Alberto Elia, Turin, Italy, ca. 1947; Model 5, power 0.2 hp at 5500 rpm

6.56 OK Cub 0.074 Aero-Model Engine

1952.

Fig. 6.57 Single-cylinder two-stroke 1.2 cm^3 aero-model engine with a two-blade propeller. Manufacturer: Herkimer Tool & Model Works, New York, USA, 1952

6.57 Microtecnica Saturno Supertigre Aero-Model Engine

1952.

Fig. 6.58 Single-cylinder two-stroke 2.5 cm^3 aero-model engine with a two-blade wooden propeller. Manufacturer: Microtecnica, Bologna, Italy, 1952; type Saturno Supertigre

6.58 Microtecnica Saturno G. 20 S Supertigre Aero-Model Engine

1952.

Fig. 6.59 Single-cylinder two-stroke 2.5 cm^3 aero-model engine. Manufacturer: Microtecnica, Bologna, Italy, 1952; type Saturno G.20 Supertigre; power 0.26 hp at 14,000 rpm

6.59 Frog 500 Aero-Model Engine

1953.

Fig. 6.60 Single-cylinder two-stroke 4.9 cm^3 aero-model engine. Manufacturer: International Model Aircraft Ltd., London, UK, 1950s

6.60 McCoy Type 29 Stunt Aero-Model Engine

1957.

Fig. 6.61 Single-cylinder two-stroke 4.9 cm^3 aero-model engine with a two-blade propeller. Manufacturer: McCoy, Testor Corporation, Rockford, Illinois, USA, 1957. type 29 Stunt; power 0.51 hp at 14,000 rpm. Kind donation of Mr. Michele La Fiura

6.61 Microtecnica Saturno G. 29 FB Engine for Marine Scale Model

1957.

Fig. 6.62 Single-cylinder two-stroke 0.95 cm^3 engine for scale motorboat models. Manufacturer: Microtecnica Saturno, Bologna, 1957; type Saturno G. 29 FB; power 0.08 hp at 15,000 rpm. Kind donation of Mr. Emanuele Vanzetti

6.62 Enya 19-VI 4006 Engine Aero-Model Engine

1970s.

Fig. 6.63 Single-cylinder two-stroke 3.24 cm^3 aero-model engine with a two-blade propeller. Manufacturer: Enya Metal Products Co., Japan, 1970s. type 19-VI 4006; power 0.28 hp at 13,000 rpm

6.63 Fixed Pitch Wooden Propeller for Aircraft Scale Model

1950s.

Fig. 6.64 Fixed pitch wooden propeller, 1950s

6.64 Rocket Engine for Scale Model

1950s.

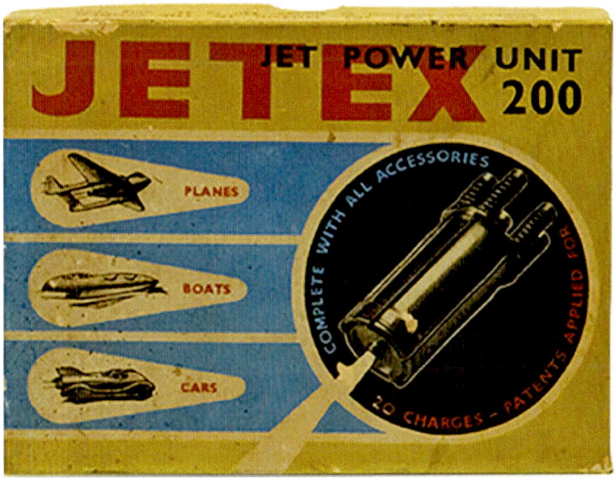

Fig. 6.65 Rocket engine, 1950s

6.65 Pulsejet Engine for Aircraft Scale Model

1950.

Fig. 6.66 Pulsejet engine for aircraft scale model, 1950

6.66 Didactic Model of Four-Stroke Spark-Ignition Engine

1950s.

Fig. 6.67 Didactic sectioned scale model of a single-cylinder four-stroke spark-ignition engine, 1950s

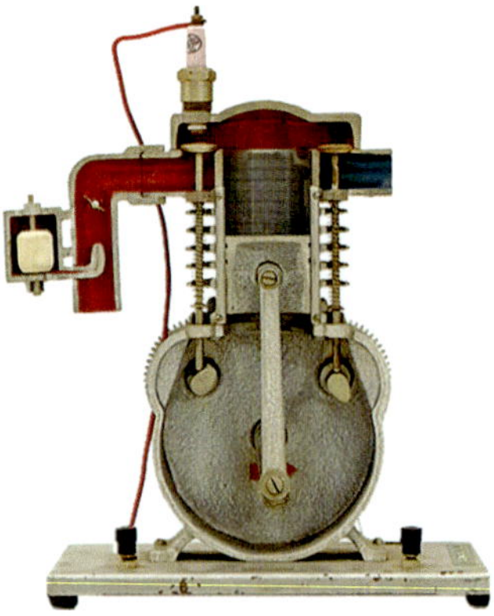

6.67 Didactic Model of a 2-stroke Diesel Engine

1950s.

Fig. 6.68 Didactic sectioned scale model of a single-cylinder 2-stroke Diesel engine, 1950s

6.68 Didactic Model of a Bosch Injection Pump for 6-cylinder Diesel Engine

1960s.

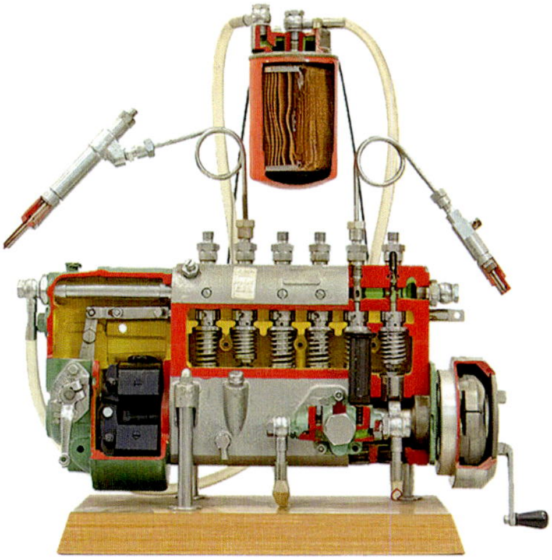

Fig. 6.69 Bosch injection pump for 6-cylinder Diesel engine, with automatic centrifugal regulator, transformed in didactic model. Manufacturer: Bosch, 1960s

6.69 Didactic Model of Wankel Rotary Engine

1960s.

Fig. 6.70 Didactic sectioned wooden model of Walkel rotary engine, 1960s

6.70 Didactic Model of an Aircraft Turbojet Engine

1950s.

Fig. 6.71 Didactic sectioned scale model of an aircraft turbojet engine with single-stage centrifugal compressor, multiple can-type combustion chambers, single-stage axial-flow turbine, and fixe geometry exhaust nozzle

6.71 Didactic Model of a Pelton Turbine with Dynamo

1930s.

Fig. 6.72 Didactic scale working model of a Pelton turbine with dynamo, 1930s

6.72 Worthington Hydraulic 4-stage Centrifugal Pump

1960s.

Fig. 6.73 Didactic sectioned four-stage centrifugal pump for industrial applications. Manufacturer: Worthington Corporation, 1960s; max. feeding pressure 59 bar; max. speed 5400 rpm

6.73 Hydraulic Single-Stage Centrifugal Pump

1986.

Fig. 6.74 Hydraulic single-stage centrifugal pump. Manufacturer: Isi, Impianti, Genova, 1986

6.74 FIAT 508 C Automotive Manual Gearbox

1937.

Fig. 6.75 Automotive 4-speed and reverse manual gearbox of a FIAT 508 C, 1937

6.75 Didactic Sectioned Model of an Automobile

1960s.

Fig. 6.76 Didactic sectioned model of an automobile. Manufacturer: C.E.A.D.A. using various component from different FIAT cars, 1960s

Chapter 7
Technical Equipment, Parts, and Accessories

Abstract The academic nature of the Museum is evident in its dedication to promoting the history of research and teaching activities, with particular regards to Machinery and Applied Mechanics. To this purpose a relevant part of its exhibit presents the evolution of motoring, highlighting technical and scientific aspects crucial to the development of various machines. Research activities form a vital bridge between the past and future, showcased through a collection of mechanical parts and scientific equipment. These instruments, dating back to the nineteenth century, were originally used for studying steam and internal combustion engines at the laboratories of the University of Palermo. Evolving alongside technological advancements, the instruments transitioned from mechanical to electronic and digital systems, offering greater precision and data processing capabilities. The collection also includes different types of dynamometric brakes for engine test bench, as well as several equipment and mechanical part and accessories, from carburettors up to modern fuel injection systems. Some items, including rare wooden propellers and cockpit instruments originate from early aviation period, adding a noteworthy historical and collectible significance even in this part of the Museum's collection.

G. Genchi, *The Museum of Engines and Mechanisms of the University of Palermo*, History of Mechanism and Machine Science 48,
https://doi.org/10.1007/978-3-031-63042-2_7

Fig. 7.1 The former Laboratory of Machinery of the University of Palermo, 1940–1950s

Fig. 7.2 The exhibit area dedicated to historical technical equipment, parts, and accessories

The academic nature of the Museum is reflected in the commitment reserved for the promotion of the history of the research and teaching activities carried out at the University of Palermo, with particular regards to Machinery and Applied Mechanics. This distinguishes the Museum's initiatives in a distinctive way, in particular the way of representing the history and evolution of motoring to the public and students, in which the technical and scientific aspects are highlighted, as well as the historical and economic ones, including those which entailed the failure of innovative ideas and particular arrangements. Research activities, which constitute an important part of the link between the past and the future of motoring, are represented through the collection of technical and scientific instruments (Figs. 7.1 and 7.2). These finds bear witness to the experimental activities carried out since 1866 at the laboratories of the Royal Application School for Engineers and Architects. The temporal evolution of the typology and technical characteristics of the devices follows hand in hand the scientific development of the various research fields, as well as the considerable technological progress carried out in the field of machinery. The oldest instruments, dating back to the second half of the nineteenth century, are those used for experimental research and didactical activities on steam engines and, subsequently, on the first internal combustion engines. Most of them concern the main physical quantities which describe the functioning of a combustion engine, such as gas or liquid pressure (Figs. 7.3, 7.4, 7.5, 7.6), rotational speed (Figs. 7.7, 7.8, 7.9, 7.10, 7.11), fluid velocity (Figs. 7.12, 7.13, 7.14, 7.15, 7.16), temperature, atmospheric pressure (Figs. 7.17 and 7.18) and gas mixture composition (Fig. 7.19). The most valuable ones are the pressure indicators used for the experimental measurement of the pressure cycles inside the engine cylinders, shown in Figs. 7.3 and 7.4 alongside with their original accessories. The latter, as well as other devices (e.g. Fig. 7.7) measure the physical quantity and, at the same time, illustrate its trend in a graph on a paper tape, as function of time or crank angle position. With the evolution of technology and the development of increasingly in-depth research, the laboratories have been equipped with increasingly advanced equipment. From purely mechanical and analogue instruments up to electric (e.g. Fig. 7.11), electronic and then digital systems. With greater precision and accuracy, the latter allow to process an increasing amount of data by the use of modern computer systems. This vast and heterogeneous collection also includes numerous mechanical components, such as carburettors, injection pumps, fuel mixture composition analyser, and various aircraft equipment (from Figs. 7.19 to 7.26). The rarest and oldest ones are part of a set coming from Germany after the end of the First World War, together with numerous engines (e.g. Fig. 3.3) and wooden aeronautical propellers (from Figs. 7.27 to 7.32). Research activity on fluid dynamics and supercharging system are well represented by air compressors (Figs. 7.33 and 7.34) and a small wind tunnel (Fig. 7.35), designed and used for research and teaching at the former Aeronautical Institute of the University of Palermo in the 1960s. Finally, different dynamometric brakes (Figs. 7.36, 7.37, 7.38, 7.39) illustrate the main types of devices commonly used in an engine test bench both to dissipate the mechanical power and to measure the torque delivered under various operating conditions.

7.1 Lehmann & Michels In-Cylinder Pressure Indicator for Piston Engine

1880–1920s.

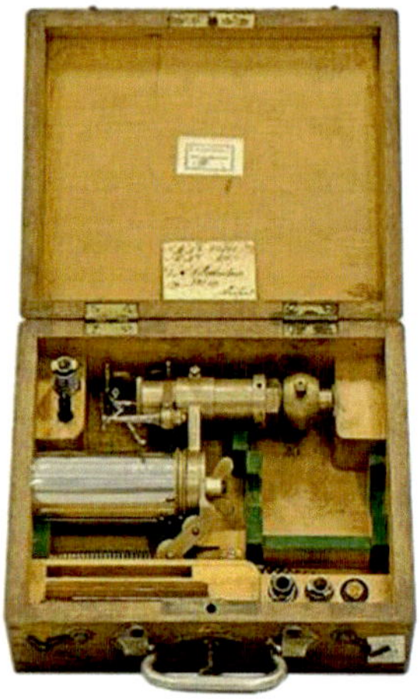

Fig. 7.3 In-cylinder pressure indicator for steam or internal combustion engine. Manufacturer: Lehmann & Michels, Hamburg, Germany, 1880–1920s

7.2 Shaeffer & Budenberg In-Cylinder Pressure Indicator for Piston Engine

1880–1920s.

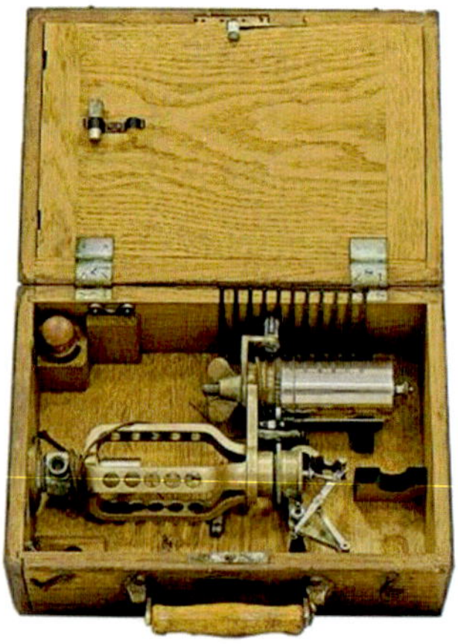

Fig. 7.4 In-cylinder pressure indicator for steam engine and internal combustion engine. Manufacturer: Shaeffer & Budenberg, Germany, 1880–1920s

7.3 Shaeffer & Budenberg Laboratory Pressure Gauge

1880–1920s.

Fig. 7.5 Laboratory pressure gauge with measurement range from 0 to 20 kg/cm^2. Manufacturer: Shaeffer & Budenberg, Germany, 1880–1920s

7.4 Shaeffer & Budenberg Quick-Connection Pressure Gauge

1880–1920s.

Fig. 7.6 Quick-connection pressure gauge with measurement range from 0 to 20 kg/cm^2. Manufacturer: Shaeffer & Budenberg, Buckau–Milan, 1880–1920s

7.5 Tachograph Recorder for Laboratory Applications

1880–1920s.

Fig. 7.7 Tachometer recorder used at the laboratory of machinery of the University of Palermo, 1880–1920s

7.6 Shaeffer & Budenberg Portable Tachometer

1880–1920s.

Fig. 7.8 Portable tachometer with three ranges of measurement, up to 2000 rpm. Manufacturer: Shaeffer & Budenberg, Buckau–Milan, 1880–1920s

7.7 Wilhem Morell Phylax Aircraft Engine Tachometer

1914–1918.

Fig. 7.9 Mechanical tachometer for aircraft engines, with measurement range from 300 to 1600 rpm. Manufacturer: Wilhelm Morell G.m.b.H., Leipzig, Germany, 1914–1918

7.8 Jaeger Automotive Engine Tachometer

1950s.

Fig. 7.10 Mechanical tachometer for automotive engines, with measurement range from 400 to 8600 rpm. Manufacturer: Jaeger, 1950s

7.9 Rabotti Electric Tachometer for Spark-Ignition Engines

1960s.

Fig. 7.11 Electric tachometer for 4- to 6-cylinder spark-ignition engines, with measurement range from 0 to 6000 rpm. Manufacturer: Rabotti S.p.A., Turin, 1960s

7.10 J. Von Petravic & Co. Aircraft Anemometer (Air Speed Indicator)

1914–1918.

Fig. 7.12 Air speed indicator used in German and Austro-Hungarian aircraft during the First World War. The device made up of a hemispherical-cup anemometer with a range of measurement from 40 to 250 km/h. Manufacturer: J. von Petravic & Co., Vienna, Austria, 1914–1918

7.11 AEG Aircraft Air Speedometer

1914–1918.

Fig. 7.13 Cockpit air speed indicator used in German and Austro-Hungarian aircraft during the First World War. The device is made up of a differential-pressure sensor for Pitot-tube system, with range of measurement from 40 to 200 km/h. Manufacturer: AEG, Germany, 1914–1918

7.12 A. Ott Kempten Liquid-Flow Speedometer

1955.

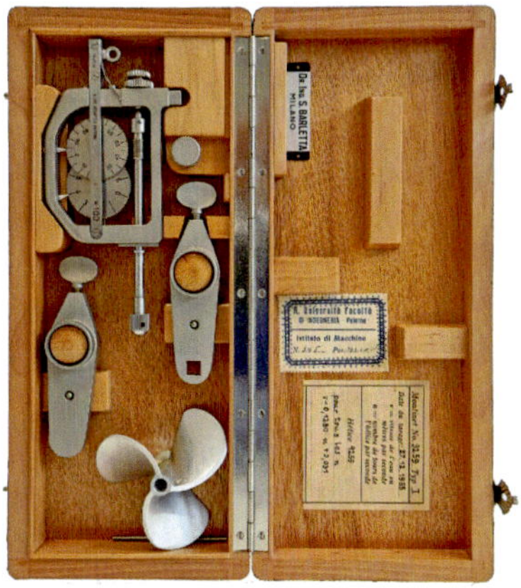

Fig. 7.14 Propeller-type liquid-flow speedometer. Manufacturer: A. Ott Kempten, Bayern, Germany, 1955

7.13 Wilhelm Lambrecht KG Anemometer

1956.

Fig. 7.15 Portable propeller-type anemometer. Manufacturer: Wilhelm Lambrecht KG, Gottingen, Germany, 1956

7.14 Anemometer for Pitot-Tube System

1950s.

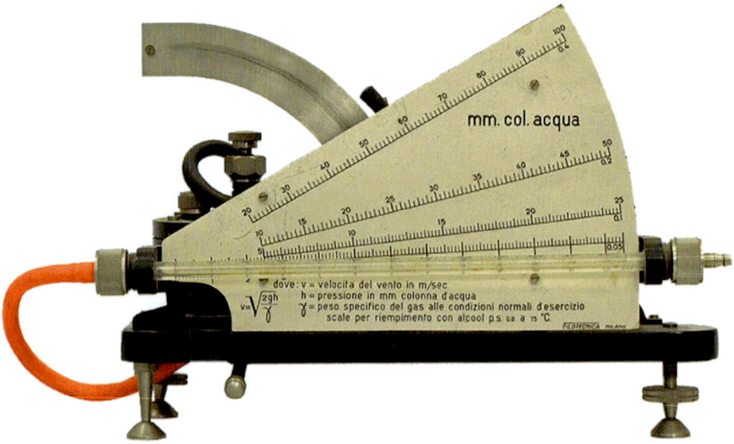

Fig. 7.16 Laboratory anemometer made up of an adjustable liquid-column differential-pressure indicator for Pitot-tube system, with three ranges of measurement up to 100 km/h

7.15 AEG Aircraft On-Board Barometer Recorder

1914–1918.

Fig. 7.17 Aircraft on-board barometer indicator-recorder used during First World War. Manufacturer: AEG, Germany, 1914–1918

7.16 Filotecnica Aircraft On-Board Barometer Recorder

1940s.

Fig. 7.18 Aircraft on-board barometer indicator-recorder, for a maximum flight altitude of 12,000 m above sea level. Manufacturer: Filotecnica, type B1/2, Milan, 1940s. Kind donation of Mr. Ottavio Drago

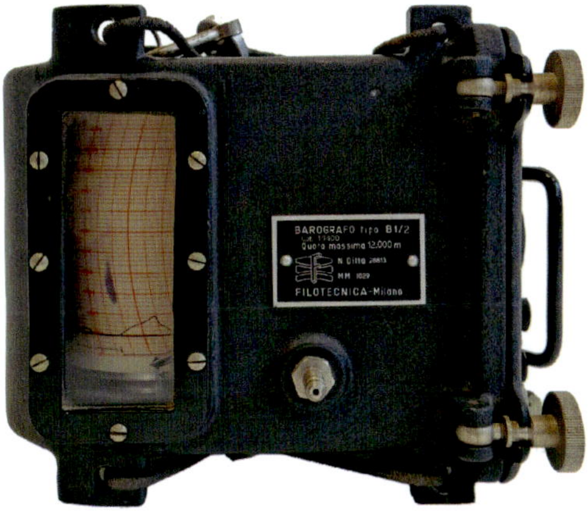

7.17 G. Lufft Aircraft Barometric Altimeter

1914–1918.

Fig. 7.19 Barometric altimeter used in German and Austro-Hungarian aircraft during the First World War, for flight altitude up to 8000 m above sea level. Manufacturer: G. Lufft, Stuttgart, Germany, 1914–1918

7.18 J. Von Petravic & Co. Aircraft Barometric Altimeter

1914–1918.

Fig. 7.20 Barometric altimeter used in German and Austro-Hungarian aircraft during the First World War, for flight altitude up to 6000 m above sea level. Manufacturer: J. von Petravic & Co., Vienna, Austria, 1914–1918

7.19 Thöne & Fiala Aircraft Pitch and Roll Indicator

1914–1918.

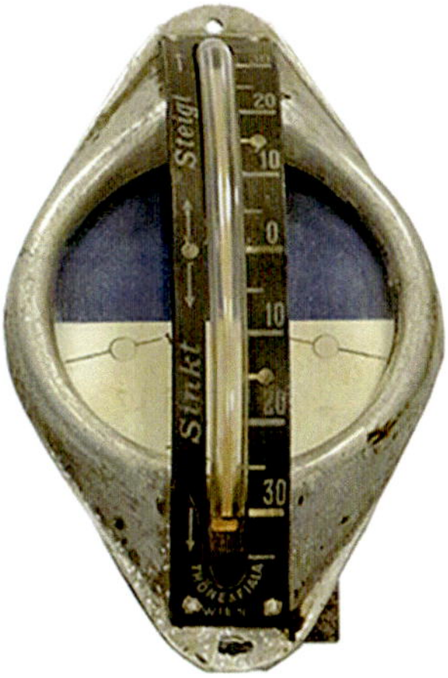

Fig. 7.21 Cockpit pitch and roll indicator (liquid-type) used in German and Austro-Hungarian aircraft during the First World War. Manufacturer Thöne & Fiala, Vienna, Austria, 1914–1918

7.20 Süss Aircraft Pitch Indicator

1914–1918.

Fig. 7.22 Cockpit pitch indicator (liquid-type) used in German and Austro-Hungarian aircraft during the First World War. Manufacturer: Süss, Budapest, Hungary, 1914–1918

7.21 Carl Bamberg Aircraft Compass—*Selenkompass*

1914–1918.

Fig. 7.23 Compass with remote control system, *Selenkompass*, used in German and Austro-Hungarian aircraft during the First World War. The device is made up of a compass with selenium-photocell sensors (left) controlling a cockpit indicator (right). Manufacturer: Carl Bamberg, Berlin Friedenau, Germany, 1914–1918

7.22 J. Von Petravic & Co. Aircraft Compass

1914–1918.

Fig. 7.24 Aircraft compass used in German and Austro-Hungarian aircraft during the First World War. Manufacturer: J. von Petravic & Co., Vienna, Austria, 1914–1918

7.23 Aircraft In-Tank Fuel Quantity Meter

1914–1918.

Fig. 7.25 In-tank fuel quantity meter used in German and Austro-Hungarian aircraft in WWI. Range of measurement from 0 to.85 L. Manufacturer: Ing. J.J. Müller., Vienna, Austria, 1914–1918

7.24 Auergesllshaft AG Oxygen Meter

1930–1940s.

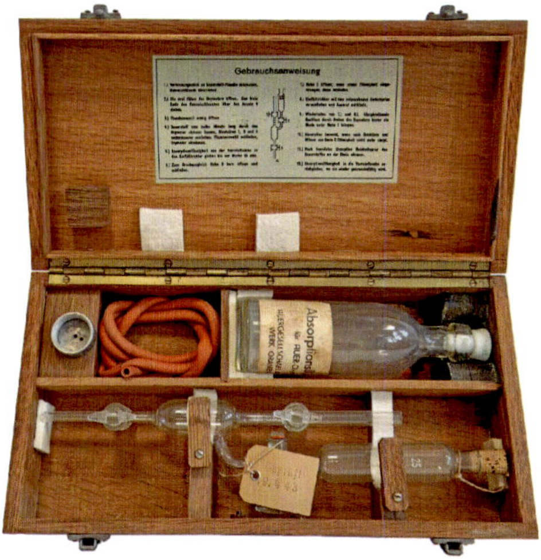

Fig. 7.26 Oxygen measurement device for assessing the air–fuel mixture composition for a combustion engine. Manufacturer: Auergesllshaft AG, Berlin, Germany, 1930–1940s

7.25 Markgraf & Tau Aircraft Propeller

1914.

Fig. 7.27 Two-blade fixed-pitch wooden propeller used by Luftstreitkräfte (German Air Force) during the First World War. Manufacturer: Markgraf & Tau—Zentral, Berlin, Germany, 1914

7.26 Franz Reschke G.M.B.H. Aircraft Propeller

1914–1918.

Fig. 7.28 Two-blade fixed-pitch wooden propeller used by Luftstreitkräfte (German Air Force) during the First World War. Manufacturer: Franz Reschke G.m.b.H., Berlin, Germany, 1914–1918

7.27 DVL Aircraft Propeller

1914–1918.

Fig. 7.29 Two-blade fixed-pitch wooden propeller used by Luftstreitkräfte (German Air Force) during the First World War. Manufacturer: Deutsche Versuchsanstalt für Luftfahrt (DVL), Berlin, Germany, 1914–1918

7.28 Wooden Aircraft Propeller

1920s.

Fig. 7.30 Two-blade fixed pitch wooden propeller for piston-engine aircraft, 1920s

7.29 Aircraft Propeller for FIAT A 20 Engine

1920s.

Fig. 7.31 Two-blade fixed pitch wooden propeller for FIAT A 20 engine, used in civil and military Italian aircrafts between the late '20s and the early '30s

7.30 Aircraft Propeller for FIAT A 50 Engine

1928–1940.

Fig. 7.32 Two-blade fixed-pitch wooden propeller for FIAT A 50 engine, used in numerous civil and military Italian aircrafts from 1928 to 1940

7.31 Compressor Impeller of a Mercedes DB 605 Aircraft Engine

1941–1945.

Fig. 7.33 Aluminium alloy impeller of the single-stage centrifugal compressor of an aircraft engine Mercedes DB 605. Manufacturer: Daimler-Benz AG, Germany, 1941–1945

7.32 Single-Stage Axial-Flow Compressor

1970s.

Fig. 7.34 Single-stage axial-flow compressor with variable-pitch inlet guide vanes, designed and used as wind tunnel fan at the former Aeronautical Institute of the University of Palermo

7.33 Wind Tunnel

1960s.

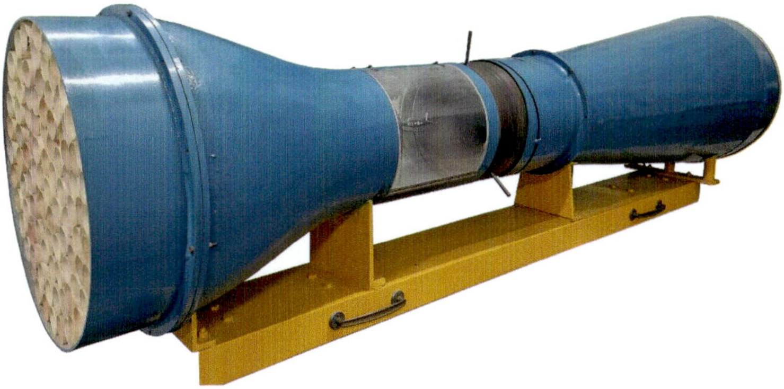

Fig. 7.35 Wind tunnel designed and used for research and didactic activities at the former Aeronautical Institute of the University of Palermo, 1960s

7.34 Ranzi Hydraulic Dynamometer

1927.

Fig. 7.36 Single-rotor water brake dynamometer used for torque measurement in engine test bench. Manufacturer: Ranzi, Legnano, Italy, 1927

7.35 Aspera Aero Dynamometer

1940s.

Fig. 7.37 Aero dynamometer for torque measurement in engine test bench. The device has a 4-blade propeller with a continuous variable-pitch system used for engine load control. Manufacturer: Aspera, Turin, Italy, 1940s. Kind donation of Mrs. Emanuela Marelli

7.36 Heenan & Froude Hydraulic Dynamometer

1962.

Fig. 7.38 Single-rotor water brake dynamometer for torque measurement in engine test bench. Manufacturer: Heenan & Froude Ltd., Worcester, United Kingdom, 1962

7.37 Heenan & Froude Electric Dynamometer

1962.

Fig. 7.39 Eddy current dynamometer brake for torque measurement in engine test bench. Manufacturer: Heenan & Froude Ltd., Worcester, United Kingdom, 1962

Appendix
Collection's Additional Items

Since the Museum foundation, and in particular after the official inauguration in 2011, its collection has considerably grown up due to numerous acquisitions both from private collectors and by means of many cooperation agreements established with various institutions. Moreover, in the last years the growth rate of the items acquired by the Museum became faster than the restoration activities carried out at the Museum's workshop, as well as the research activities operated for the first ones, exhibit design and enhancement. For these reasons, despite their technical and historical importance, many of the items are not yet restored and cannot be displayed before completing certain necessary operations at least, ranging from a preliminary protection treatment up to the construction of supports suitable for the exhibit at the Museum. Some of these acquisitions include very rare engines, unique prototypes, as well as the VM-Maserati engine just donated during the final revision of this book. They are shortly reported in this final appendix and will be of course the object of the next development activities of the Museum (Fig. A.1).

Stationary engines

Crossley O 116, donation of Mrs. Luigina Bongiovanni

Lombardini LDA 100, donation of Mr. Nello Salsapariglia

Lombardini LDW 903

Marine engines

Marine engine G. Carraro[i]

OM 002D-65-I2[i]

Aircraft engines

FIAT A 12 bis (partially sectioned)

BMW VI[i]

G. Genchi, *The Museum of Engines and Mechanisms of the University of Palermo*, History of Mechanism and Machine Science 48,
https://doi.org/10.1007/978-3-031-63042-2

Fig. A.1 The Museum's deposit in which various collection's items, such as those from recent donations, are preserved up to their restoration and exhibit

FIAT A 14 (partially sectioned)[i]

FIAT A15 R[i]

FIAT A 18[i]

FIAT A 19[i]

FIAT A 26[i]

FIAT A 50 (mounted as exploded assembly)

Isotta Fraschini Asso 500[i]

Isotta Fraschini K14[i]

FIAT A 33 RC 35[i]

FIAT A 55[i]

FIAT A 76 RC 38[i]

FIAT A 82 RC 42[i]

FIAT A 38 RC 15-45 (single-propeller prototype)[i]

FIAT A 38 RC 15-45 (two counter rotating propeller prototype)[i]

Piaggio P XVI[i]

Daimler Benz DB 603[i]

BRAMO 323 Fafnir[i]

Allison V1710-113[i]

Lycoming VO 435[i]

De Havilland Ghost[i]

Automotive and motorcycle engines

FIAT 109, donation of Mr. Gianpaolo La Colla

FIAT 108—*Balilla*, donation of Mr. Gianpaolo La Colla

Caproni Capriolo 125 (motorcycle powertrain), donation of Mr. Emanuele Vanzetti

FIAT 120—Nuova 500 Giardiniera

FIAT 305—FIAT Campagnola Diesel

Jaguar XK 4.2, donation of Mr. Elio Caramazza

Alfa Romeo head—*GTA 1300 Junior*, donation of Mr. G. Pucci di Benisichi

Honda CBX 1000 (motorcycle powertrain), donation of Mr. Eugenio Bordonali

Kawasaki GPz 900 R (motorcycle powertrain), courtesy of Mr. Roberto Licari

FIAT 156c.046—*1100 FIRE i.e.*, donation of Mr. Vito Veninata

Alfa Romeo AR 06166—*2.0 Twin Spark*

Honda Sport 4 T 50 cc motorcycle

Moto Guzzi Stornello 160 cc motorcycle

VM-Maserati 3.0 V6, donation of Mr. Bartolo Tasca, Cronos s.r.l.

Didactic mechanisms and models

МАРЗ 2,5 Д aero model engine, donation of Prof. Baldassare Ezio Adelfio

Davies Charlton Ltd., Dart, aero model engine, donation of Prof. B.E. Adelfio

Aurora Model MFG. Co., K75, aero model engine, donation of Prof. B.E. Adelfio

Davies Charlton Ltd., Merlin, aero model engine, donation of Prof. B.E. Adelfio

Regner *Lilly* steam engine for scale models, donation of Prof. Baldassare Ezio Adelfio

Technical equipment, parts, accessories

Benz & Cie. Bz.IV connecting rod and liquid cooling centrifugal pump

Mercedes D.IVa 4-piston reciprocating oil pump and engine valves

Maybach Mb.IV connecting rod

Noris, Bosch et al. starting high-voltage magnetos

Siemens & Halske Sh.III cylinder assembly with valvetrain

Curtis-Wright *Cyclone*-type cylinder assembly with piston (for radial engine)

Alfa Romeo 4-speed gearbox—*Giulia Sprint/Spider*

Alfa Romeo Giulia GTA engine head, donation of Mr. Gianfranco Pucci di Benisichi

Bosch mechanical petrol injection pump (4-cylinder engine model)

FIAT 525* 4-speed gearbox

Van Doorne CVT gearbox for Ford automobiles

Colombo inverter gearbox

Zenit H3-5 epicyclic gearing inverter gearbox

Aspera aero dynamometer, type M.T. 7/A

Ranzi hydraulic dynamometer, type D. 1/75, with control unit

Zöllner & Co. hydraulic dynamometer, type 3N16

Wincharger Corporation model 670 (wind turbine complete assembly)

Note

i. Entrusted by Aeronautica Militare (Italian Air Force) for restoration and exhibit.

References

Beccari A, Caputo C (1987) Motori termici volumetrici. UTET, Turin

Beccari A (1993) Macchine. CLUT, Turin

Barizza S (2003) La fonderia Neville a San Rocco. Un'industria del ferro a Venezia nella seconda metà dell'Ottocento. In: Storia di Venezia, Rivista I, pp 129–139. Firenze University Press, Florence. ISSN 1724–7446

Bollettino Tecnico (1972) FIAT - Ansaldo - C.R.D.A., Grandi Motori Trieste

Bollettino Tecnico (1948–1971) FIAT, Stabilimento Grandi Motori, Turin

Ceccarelli M, Sorge F, Genchi G (2016) Elia Ovazza, Professor of TMM in Palermo around the end of the 19th century. In: Essays on the history of mechanical engineering, pp 47–63. Springer

Fondo motori aeronautici, Centro Storico FIAT. Stellantis Group, Turin

Compagnino L (2021) La storia che si muove. City Rivista Palermo

De Montis M, Genchi G (2012) Cilindri nel vento, la strana e breve storia dei motori rotative. Autotecnica (377)

De Virgilio G, Boscarelli L, Morello L, De Virgilio L, Monno M, Limone S (2015) Lancia, uomini, tecnica, vittorie. Il Cammello, Turin

Douglas CE (2022) La sfida per la Potenza. Gli archivi ritrovati, Editore Paolo Miana, Gallarate. ISBN: 9788894609813

Douglas CE (2020) The secret horsepower race. Mortons Media Group. ISBN: 1911658506

Filippi F (1965) Appunti per una storia del motore aeronautico in Italia. In: Atti e Rassegna Tecnica della Società Ingegneri e Architetti in Torino, anno 19 n 10

Filippi F (1983) Dall'elica al getto. Associazione Industriali Metallurgici Affini, Editions EDA, Turin

Flight International (1909–2022) Reed Business Information Ltd

Gabrielli G (1982) Una vita per l'aviazione. Bompiani, Milan

Garuffa E (1929) I motori veloci a combustione interna e le loro applicazioni, 2nd edn. Unione Tipografico-Editrice Torinese, Turin

Genchi G, Sorge F (2012) the rotary aero engine from 1908 to 1918, In: Koetsier T, Ceccarelli M (eds), History of mechanism and machine science, pp 349–362. Springer, Dordrecht. ISBN 9789400741317

Genchi G (2016) Il Museo Storico dei Motori e dei Meccanismi dell'Università di Palermo. In: Scuola Officina, Rivista del Museo del Patrimonio Industriale di Bologna (1)

Genchi G (2017) The collection of engines at the museum of engines and mechanisms, University of Palermo. American Society of Mechanical Engineers, ASME Landmark (264)

G. Genchi, *The Museum of Engines and Mechanisms of the University of Palermo*, History of Mechanism and Machine Science 48, https://doi.org/10.1007/978-3-031-63042-2

Genchi G (2017) I motori e il mito della velocità. In: Il ruggito della velocità: miti e modernità della Targa Florio motociclistica, pp 54–55 and 161–165. Regione Siciliana, Assessorato dei beni culturali e dell'identità siciliana, Palermo. ISBN: 978-88-6164-478-6
Genchi G (2018) Il Sistema Museale di Ateneo, tutela e sviluppo del patrimonio storico dell'Università di Palermo. Quotidiano di Sicilia
Genchi G, Bellomo G (2018) Il Museo Storico dei Motori e dei Meccanismi di Palermo. Nuova Museologia
Giacosa D (1979) I miei 40 anni di progettazione alla FIAT. Automobilia
Giacosa D (1941) Motori endotermici. Hoepli, Milan
Giger H (1996) Kolben-Flugmotoren. Motorbuch Verlag. ISBN-10 3613010895
Grammauta F, Caleca L (1986) La centrale elettrica A. Volta di Palermo: vicende storiche, analisi tecnico-funzionale ed ipotesi di una riutilizzazione nell'ambito del contesto urbanistico del molo trapezoidale. Graduation thesis, University of Palermo
Greco F (2018) Il motore aeronautico FIAT. Gli Archivi Ritrovati, Gallarate. ISBN 8894244040
Gunston B (2006) Development of piston aero engines. 2nd edn. Patrick Stephens Limited, Sparkford
Gunston B (2007) World Encyclopedia of Aero Engines: From the pioneers to the present day. 5th edn. Sutton Publishing. ISBN-10 075094479X
Illustrato FIAT (1961), FIAT, Turin
Jane's Aero-Engines World (1919–2022). Jane's Information Group
Katz H (1928) Neuzeitliche Flugmotoren. R.C. Schmidt, Berlin
La Mantia F P (2006) Contributi per una storia della Facoltà di Ingegneria di Palermo. Edizioni Fotograf, Palermo
Lanzara G, Rubino M, De Gregorio C (1962) Motore a pistoni rotanti: progetto e realizzazione sperimentale. Graduation thesis, University of Palermo
Lawler A (2011) A grad student in Italy salvage Germany's rarest World War I airplane engines. AIR&SPACE, Smithsonian Institution
Malizia N (2005) FIAT G. 59. IBN, Rome. ISBN 8875650187
Marchi O (1980) Catalogo motori del Museo Storico dell'Aeronautica Militare. Editrice Pàtron, Bologna
Marchi O, Zardo V (1980) Catalogo velivoli del Museo Storico dell'Aeronautica Militare. Editrice Pàtron, Bologna
Martorelli GF (1910) I motori marini, 4th edn. Società Tipografico-Editrice Nazionale, Turin
Miana P, De Montis M (2015) Il FIAT G.59 rinasce al Museo Motori di Palermo. VS Aviation
Monastero R, Genchi G (2012) Macchine: Energia e Storia. In: Gerbino A (ed.) Almanacco di cultura, pp 146–152. Plumelia Edizioni, Bagheria. ISBN 9788889876206
Monastero R, Genchi G (2016) The museum of engines and mechanisms. more than a century of history of technology. In: Sorge F, Genchi G (eds), Essays on the history of mechanical engineering, pp 201–225. Springer. https://doi.org/10.1007/978-3-319-22680-4
Morello L (2011) FIAT, Storie di innovazione tecnologica nelle automobile. Associazione Tecnica dell'Automobile. ISBN 8890581514
Morello L, Genta G (2013) L'automobile: evoluzione di una tecnologia. ASI Service, Turin
Morello L, Genta G. Grandi M (2017) La più veloce. ASI Service, Turin
Morello L, Bellomo G (2018) All'improvviso un motore Isotta Fraschini. La Manovella
Morello L (2014) Lancia, Storie di innovazione tecnologica nelle automobile. FIAT Group Marketing & Corporate Communication S.p.A.
Morello L (2016) L'automobile a vapore, storia e prospettive. ASI Service, Turin
Morello L (2021) FIAT's contribution to the evolution of automobile technology. Registro Storico FIAT
Nauhm A (1987) The rotary aero engine. The Stationery Office/Tso. ISBN 0112904521
Sorge F, Genchi G (Eds) Essays on the history of mechanical engineering. Springer International Publishing Switzerland. ISBN: 978-3-319-22679-8

Valdonio G C (2019) Frecce, saette, folgori e veltri. Storia critica dei Caccia italiani della Seconda guerra Mondiale. Rivista Aeronautica. ISBN-10 8888180397

von Gersdorff K, Schubert H, Hebert S (2007), Flugmotoren und Strahltriebwerke. 4th edn. Bernard und Graefe, Bonn. ISBN: 13 978-3763761289

Zampicinini F (1994) La collezione "Antonio Capetti" di motori alternativi per aeromobili. Collana Collezioni 1, Politecnico di Torino, Turin

Zima S, Ficht R (2010) Ungewohnliche motoren. 3rd edn. Vogel Communications Group GmbH & Co. KG. ISBN: 13 978-3834331403

Index

G. Genchi, *The Museum of Engines and Mechanisms of the University of Palermo*, History of Mechanism and Machine Science 48,
https://doi.org/10.1007/978-3-031-63042-2

GPSR Compliance

The European Union's (EU) General Product Safety Regulation (GPSR) is a set of rules that requires consumer products to be safe and our obligations to ensure this.

If you have any concerns about our products, you can contact us on ProductSafety@springernature.com

In case Publisher is established outside the EU, the EU authorized representative is:

Springer Nature Customer Service Center GmbH
Europaplatz 3
69115 Heidelberg, Germany

Batch number: 10756376

Printed by Printforce, the Netherlands